Lecture Notes in Mathematics

Edited by A. Dold and B. Eckmann

536

Wolfgang M. Schmidt

Equations over Finite Fields An Elementary Approach

Springer-Verlag
Berlin · Heidelberg · New York 1976

Author

Wolfgang M. Schmidt
Department of Mathematics
University of Colorado
Boulder, Colo., 80309/USA

Library of Congress Cataloging in Publication Data

Schmidt, Wolfgang M
   Equations over finite fields.

   (Lecture notes in mathematics ; 536)
   Bibliography: p.
   1. Diophantine analysis. 2. Modular fields.
I. Title. II. Series: Lecture notes in mathe-
matics (Berlin) ; 536.
QA3.L28   vol.536   [QA242]   510'.8s [512.9'4]
                                76-26612

AMS Subject Classifications (1970): 10 A 10, 10 B 15, 10 G 05, 12 C 25, 14 G 15

ISBN 3-540-07855-X Springer-Verlag Berlin · Heidelberg · New York
ISBN 0-387-07855-X Springer-Verlag New York · Heidelberg · Berlin

© by Springer-Verlag Berlin · Heidelberg 1976

Printed in Germany

Printing and binding: Beltz Offsetdruck, Hemsbach/Bergstr.

Preface

These Lecture Notes were prepared from notes taken by M. Ratliff and K. Spackman of lectures given at the University of Colorado.

I have tried to present a proof as simple as possible of Weil's theorem on curves over finite fields. The notions of "simple" or "elementary" have different interpretations, but I believe that for a reader who is unfamiliar with algebraic geometry, perhaps even with algebraic functions in one variable, the simplest method is the one which originated with Stepanov. Hence it is this method which I follow.

The length of these Notes is perhaps shocking. However, it should be noted that only Chapters I and III deal with Weil's theorem. Furthermore, the style is (I believe) leisurely, and several results are proved in more than one way. I start in Chapter I with the simplest case, i.e., with curves $y^d = f(x)$. At first I do the simplest subcase, i.e., the case when the field is the prime field and when d is coprime to the degree of f. This special case is now so easy that it could be presented to undergraduates. The general equation $f(x,y) = 0$ is taken up only in Chapter III, but a reader in a hurry could start there. The second chapter, on character sums and exponential sums, is included at such an early stage because of the many applications in number theory. Chapters IV, V and VI deal with equations in an arbitrary number of variables.

Possible sequences are chapters

I by itself, or

I, III for Weil's theorem, or

I.1,III for a reader who is in a hurry, or

I, II for character sums and exponential sums, or

I, II, IV, or

I, III, IV.3 and V .

Originally I had planned to include Bombieri's version of the Stepanov method. I did include it in my lectures at the University of Colorado, but I first had to prove the Riemann-Roch Theorem and basic properties of the zeta function of a curve. A proof of these basic properties in the Lecture Notes would have made these unduly long, while their omission would have made the Bombieri version not self complete. Hence I decided after some hesitation to exclude this version from the Notes.

Recently Deligne proved far reaching generalizations of Weil's theorem to non-singular equations in several variables, thereby confirming conjectures of Weil. It is to be noted, however, that Deligne's proof rests on an assertion of Grothendieck concerning a certain fixed point theorem. To the best of my knowledge, a proof of this fixed point theorem has not appeared in print yet. It is perhaps needless to say that at present there is no elementary approach to such a generalization of Weil's theorem. But it is to be hoped that some day such an approach will become available, at least for those cases which are used most often in analytic number theory.

November, 1975 W. M. Schmidt

Notation

F^* is the multiplicative group of a field F .

$\bar{F}$ is the algebraic closure of a field F .

F^n is the product $F \times \dots \times F$, i.e., the set of n-tuples $(x_1,\dots,x_n)$ with $x_i \in F$ $(i = 1,\dots,n)$.

$[F_1 : F_2]$ denotes the degree of a field extension $F_1 \supseteq F_2$.

$\mathfrak{T}$ denotes the trace and $\mathfrak{N}$ the norm.

F_q will denote the finite field with q elements.

p will be the characteristic.

$\mathbb{Q}$ is the field of rational numbers,

$\mathbb{R}$ the field of reals,

$\mathbb{C}$ the field of complex numbers,

$\mathbb{Z}$ the ring of (rational) integers.

$\cong$ denotes isomorphism of fields or groups.

Quite often, $x,y,z\dots$ will be elements which lie in a ground field or are algebraic over a ground field, $X,Y,Z,\dots$ will be variables, i.e., will be algebraically independent over a ground field, and $\mathfrak{X}$, $\mathfrak{Y},\dots$ will be algebraic functions, i.e., they will be algebraically dependent on some of $X,Y,\dots$. Thus $f(X_1,\dots,X_n)$ is a polynomial, and $f(x_1,\dots,x_n)$ is the value of this polynomial at $(x_1,\dots,x_n)$.

$F(x)$ or $F(X)$ or $F(X,Y)$ or $F(X,\mathfrak{Y})$, or similar, will be the field obtained by adjoining x or X or X,Y or $X,\mathfrak{Y}$ to a ground field F . Thus $F(X)$ is the field of rational functions in a variable X with coefficients in F . $R[X]$ denotes the ring of polynomials in X with coefficients in the ring R .

If a,b are in Z, we write $a|b$ (or $a \nmid b$) if a does (or does not) divide b. Occasionally we shall write $d|q-1$ instead of the more proper notation $d|(q-1)$. Again, we shall write $f(X)|g(X)$ if the polynomial $f(X)$ divides $g(X)$. Further $(f(X))$ (or $(f(X),g(X))$) will be the ideal generated by $f(X)$ (or by $f(X)$ and $g(X)$).

$|\omega|$ denotes the number of elements of a finite set ω. Given sets $A \subseteq B$, the set theoretic difference is denoted by $B \sim A$.

Table of Contents

Introduction

Gauss (1801) made an extensive study of quadratic congruences modulo a prime p. He also obtained the number of solutions of the cubic congruence

$$ax^3 - by^3 \equiv 1 \pmod{p}$$

for primes $p = 3n+1$, and of the quartic congruence

$$ax^4 - by^4 \equiv 1 \pmod{p}$$

for primes $p = 4n+1$. He studied the congruence

$$ax^4 - by^2 \equiv 1 \pmod{p}$$

for arbitrary primes p.

Artin (1924) considered the congruence $y^2 \equiv f(x) \pmod{p}$, where $f(X)$ is a polynomial whose leading coefficient is not divisible by p and which has no multiple factors modulo p, and made the following conjecture: The number N of solutions satisfies

$$|N - p| \leq 2\sqrt{p} \quad \text{if} \quad \deg f = 3 ,$$

$$|N+1 - p| \leq 2\sqrt{p} \quad \text{if} \quad \deg f = 4 .$$

This conjecture was proved by Hasse (1936 b,c.). In fact, let F_q be the finite field with q elements, and let N be the number of solutions $(x,y) \in F_q^2$ of the equation $y^2 = f(x)$, where $f(X)$ is a polynomial with coefficients in F_q and with distinct roots. Then

$$|N - q| \leq 2\sqrt{q} \quad \text{if} \quad \deg f = 3 ,$$

$$|N+1 - q| \leq 2\sqrt{q} \quad \text{if} \quad \deg f = 4 .$$

Suppose $f(X,Y)$ is a polynomial of total degree d, with coefficients in F_q and with N zeros (x,y) with coordinates in F_q. Suppose $f(X,Y)$ is absolutely irreducible, i.e., irreducible not only over F_q, but also over every algebraic extension thereof.

Weil (1940,1948a)[†] proved the famous theorem (the "Riemann Hypothesis for Curves over Finite Fields") that

(1) $$|N - q| \leq 2g\sqrt{q} + c_1(d)$$

where g is the "genus" of the curve $f(x,y) = 0$ and where $c_1(d)$ is a constant depending on d. It can be shown that $g \leq \frac{1}{2}(d-1)(d-2)$, hence that

$$|N - q| \leq (d-1)(d-2)\sqrt{q} + c_1(d) .$$

Weil's proof depends on algebraic geometry, in particular on Castelnuovo's inequality. A somewhat simpler proof was given by Roquette (1953); see also Lang (1961), Eichler (1963).

More recently, Stepanov (1969, 1970, 1971, 1972a, 1972b, 1974) gave a new proof of special cases of Weil's result which does not depend on algebraic geometry, but which is related to Thue's (1908) method in diophantine approximation. This method consists in the construction of a polynomial in one variable with rather many zeros. The construction is by the method of undetermined coefficients.

In particular, Stepanov proved that

(2) $$|N - q| \leq c_2(d)\sqrt{q}$$

if $f(X,Y)$ is of some special type, for instance if

$$f(X,Y) = Y^d - f(X)$$

where d and the degree of f are coprime. Later Bombieri (1973) and Schmidt proved (2) for absolutely irreducible $f(X,Y)$ by the Thue - Stepanov method. It follows from the theory of the zeta function that (2) implies (1).

In these Lectures we shall prove (2) by the Stepanov method.

[†]The 1940 paper is only an announcement.

I. Equations $y^d = f(x)$ and $y^q - y = f(x)$.

References: Stepanov (1969, 1970, 1971, 1972a), Mitkin (1972), Stark (1973).

§ 1. Finite Fields (Galois fields).

Let F be any field. There is a smallest subfield $k \subseteq F$ (the intersection of all subfields of F), called the prime subfield of F, and either $k = \mathbb{Q}$ or $k = F_p$, the integers modulo a prime p. In the first case F is of characteristic 0, in the second case of characteristic p. In the case when F is finite, $k = F_p$, and $[F : F_p]$ is finite. If, say, $[F : F_p] = \kappa$, then $|F| = p^\kappa$. Hence if F_q is a field with q elements, then $q = p^\kappa$, p prime.

Let F_q be a finite field and let F_q^* be the multiplicative group of F_q. Then $|F_q^*| = q - 1$. If $x \in F_q^*$, then $x^{q-1} = 1$; hence, for $x \in F_q$, we have $x^q - x = 0$. Therefore, $X^q - X = \prod_{x \in F_q} (X - x)$. So F_q is the splitting field of $X^q - X$ over F_p, and F_q is a normal extension of F_p. Moreover, as a splitting field, F_q is unique up to isomorphisms.

Conversely, let F be the splitting field of $X^q - X$ over F_p, where $q = p^\kappa$. Let $x_1, \ldots, x_q$ be the roots of this polynomial in F. These roots are distinct since the derivative $D(X^q - X) = -1 \neq 0$. Now $x_i + x_j$ is a root of $X^q - X$, since,

$$(x_i + x_j)^q - (x_i + x_j) = x_i^q + x_j^q - x_i - x_j = 0 ,$$

and similarly for $x_i - x_j$. Also $x_i x_j$ is a root, since

$$(x_i x_j)^q = x_i^q x_j^q = x_i x_j ,$$

and similarly x_i/x_j is a root if $x_j \neq 0$. These roots clearly form a field, so, in fact, $F = \{x_1, x_2, \ldots, x_q\}$. Thus a field with q elements does exist.

Considering the above, we have:

THEOREM 1A. *If F_q is a finite field of order q, then $q = p^K$, p prime. For every such q, there exists exactly one field F_q. This field is the splitting field of $X^q - X$ over F_p, and all of its elements are roots of $X^q - X$.*

THEOREM 1B. *The multiplicative group F_q^* is cyclic.*

For the proof of this theorem we need

LEMMA 1C. *Let G be a finite group of order d. Suppose for every divisor e of d, there are at most e elements $x \in G$ with $x^e = 1$. Then G is cyclic.*

The theorem follows immediately, since $X^e - 1$ has at most e roots in F_q^*. It only remains to give a

Proof of Lemma 1C. Every element of G is of some order e, where $e|d$. Let $\psi(e)$ be the number of elements of G whose order is e. Either $\psi(e) = 0$ or $\psi(e) \neq 0$. Suppose $\psi(e) \neq 0$, and let $y \in G$ have order e. Then the elements $y, y^2, \ldots, y^e = 1$ are distinct and all satisfy $x^e = 1$. Since there are e of these elements, by hypothesis there can be no other elements $x \in G$ satisfying $x^e = 1$.

Now let $z \in G$ be any element of order e; then $z = y^i$ $(1 \le i \le e)$. Notice that $z = y^i$ has order e precisely if $(i,e) = 1$. Hence $\psi(e) = \varphi(e)$, where φ is the Euler φ-function. So, in general, $\psi(e) \le \varphi(e)$, taking into account the possibility that $\psi(e) = 0$. But

$$d = \sum_{e|d} \psi(e) \le \sum_{e|d} \varphi(e) = d .$$

Hence, for every divisor e of d, $\psi(e) = \varphi(e)$; in particular, $\psi(d) = \varphi(d) \neq 0$. That is, there exists an element of order d; hence, G is cyclic.

COROLLARY 1D. *Let* $q = p^{\kappa}$. *Then* $F_q = F_p(x)$ *for some* x.

Proof. Let x be a generator of F_q^*.

Let $F_q \subseteq F_r$ be finite fields; then $r = q^{\varkappa}$. Consider the mapping $\omega : F_r \to F_r$ such that $\omega(x) = x^q$. This mapping is one-one. For suppose $x^q = y^q$, then

$$0 = x^q - y^q = (x - y)^q ,$$

whence $x - y = 0$ and $x = y$. The mapping ω is then one-one of a finite set to itself, hence is onto. Moreover, ω is an automorphism of F_r, since

$$\omega(x + y) = (x + y)^q = x^q + y^q = \omega(x) + \omega(y)$$

and $$\omega(xy) = (xy)^q = x^q y^q = \omega(x)\,\omega(y) .$$

In fact, ω is an automorphism of "F_r over F_q" (leaving F_q fixed), since if $x \in F_q$, $\omega(x) = x^q = x$. In other words,

ω is a member of the Galois group of F_r over F_q. The map ω is called the "Frobenius automorphism".

If $r = q^K$, then $1, \omega, \omega^2, \ldots, \omega^{K-1}$ are automorphisms of F_r over F_q, and they are distinct because if

$$\omega^i = \omega^j \qquad (0 \le i,\ j \le K - 1),$$

then $$\omega^i(x) = \omega^j(x) \qquad \text{for all } x \in F_r,$$

$$x^{q^i} = x^{q^j} \qquad \text{for all } x \in F_r,$$

so $$x^{q^i} - x^{q^j} = 0 \qquad \text{for all } x \in F_r.$$

But the degree of the polynomial $X^{q^i} - X^{q^j}$ is less than $q^K = r$, so the above cannot hold identically for all $x \in F_r$, unless $X^{q^i} - X^{q^j}$ is identically zero and $i = j$. Since the order of the Galois group is K, these are the only automorphisms of F_r over F_q. We have shown:

THEOREM 1E. Every automorphism of F_r over F_q is of the form ω^i $(0 \le i \le K - 1)$, where $\omega(x) = x^q$. That is, the Galois group of F_r over F_q is cyclic with generator ω.

Recall that the trace of an element is the sum of its conjugates. For the case $F_q \subseteq F_r$, the trace of an element $x \in F_r$ is

$$\mathcal{T}(x) = x + x^q + x^{q^2} + \ldots + x^{q^{K-1}}.$$

LEMMA 1F. Let $x \in F_r$, with $F_q \subseteq F_r$. Then the following three conditions are equivalent:

(i) $\mathfrak{T}(x) = 0$.

(ii) There exists $y \in F_r$ with $x = y^q - y$.

(iii) There exist precisely q elements $y \in F_r$ with $x = y^q - y$.

Proof: Exercise.

Now let K be any field of characteristic p. Then the mapping $\omega : x \to x^p$ is an endomorphism of K. However, in this case, ω need not be onto.

Example: Let $K = F_p(X)$, p prime. Then

$$\omega(a_0 + a_1X + \dots + a_tX^t) = a_0 + a_1X^p + \dots + a_tX^{tp} .$$

Here
$$\omega(F_p(X)) = F_p(X^p) .$$

It is clear, however, that ω is onto whenever K is algebraically closed.

Let $k[X]$ denote the ring of polynomials over k. Let D be the differentiation operator defined as usual:

$$D(a_0 + a_1X + \dots + a_tX^t) = a_1 + 2a_2X + \dots + ta_tX^{t-1} .$$

THEOREM 1G. Let k be a field of characteristic p, and let M be an integer, $M \le p$. Suppose $a(X) \in k[X]$ and for some $x \in k$,

$$0 = a(x) = Da(x) = D^2a(x) = \dots = D^{M-1}a(x) .$$

Then $a(X)$ has a zero at x of order M; i.e., $(X - x)^M$ divides $a(X)$, or in symbols, $(X - x)^M \mid a(X)$.

Proof: Write

$$a(X) = c_0 + c_1(X-x) + c_2(X-x)^2 + \dots + c_t(X-x)^t .$$

Then, $D^\ell a(X) = \ell! \left[c_\ell + \binom{\ell+1}{\ell} c_{\ell+1}(X-x) + \dots + \binom{t}{\ell} c_t (X-x)^{t-\ell}\right]$.

Substituting x, for $0 \le \ell \le M - 1$, we have

$$0 = \ell!\, c_\ell .$$

But $\ell \le M - 1 < p$, so $\ell! \ne 0$ in k. Hence $c_\ell = 0$, $0 \le \ell \le M - 1$. It now follows that $(X - x)^M$ divides $a(X)$.

Remark: The condition $M \le p$ is essential in the above theorem. For example, consider $a(X) = X^p$. All derivatives vanish at $x = 0$, yet $a(X)$ has a zero only of order p at $x = 0$.

§2. Equations $y^d = f(x)$.

Special cases of these equations are equations

$$y^2 = f(x) ,$$

where $f(X)$ has distinct roots and is of degree 3 or 4. Such equations are called elliptic equations. Equations of the type

$$y^2 = f(x) ,$$

with an arbitrary polynomial $f(X)$ are called hyperelliptic equations. We now are going to make some heuristic arguments on hyperelliptic equations.

If $q = 2^{\kappa}$, the mapping $y \to y^2$ is the Frobenius automorphism of F_q, so as y takes on all values in F_q, so does y^2, and conversely. It is then clear that the number of solutions of $y^2 = f(x)$ is equal to the number of solutions of $y = f(x)$, which is q. On the other hand if q is odd, the number of squares in F_q^* is $\frac{q-1}{2}$, because

if $\quad F_q^* = \{g, g^2, g^3, \ldots, g^{q-1} = 1\}$,

then $\quad (F_q^*)^2 = \{g^2, g^4, \ldots, g^{q-1}\}$ and $|(F_q^*)^2| = \frac{q-1}{2}$.

One might, therefore, expect that for about half of the elements $x \in F_q$, $f(x)$ will be in $(F_q^*)^2$. For such an x, there are two values, namely y and $-y$, with $y^2 = f(x)$. So again we might expect roughly $2 \cdot \frac{1}{2} q = q$ solutions (x,y) of our equation.

Let us now refine our intuitions by way of two examples.

<u>Example 1.</u> Consider the solutions $(x,y) \in F_q \times F_q$ of the equation

$$y^2 = x^4 + 2x^2 + 1 ,$$

or

$$(y - (x^2 + 1))(y + (x^2 + 1)) = 0 .$$

Then either

$$y = x^2 + 1$$

or

$$y = -(x^2 + 1) .$$

So there are approximately $2q$ solutions to this equation. The problem appears to arise because $Y^2 - f(X)$ is reducible over F_q.

Example 2. Consider $y^2 = 2x^4 + 4x^2 + 2$ over F_3. Then

$$(y - \sqrt{2}\,(x^2 + 1))\,(y + \sqrt{2}\,(x^2 + 1)) = 0 .$$

This factorization, of course, cannot occur in F_3, since 2 is not a square in F_3. However, after adjoining a root of $X^2 - 2$ to F_3 (extending to F_9), the above factorization can be made. That is, the polynomial $Y^2 - 2X^4 - 4X^2 - 2$ is irreducible over F_3, but not absolutely irreducible. Now if either

$$\begin{array}{ll} & y - \sqrt{2}\,(x^2 + 1) = 0 \\ \text{or} & y + \sqrt{2}\,(x^2 + 1) = 0 , \end{array} \qquad (x,y \in F_3)$$

then since $\{1, \sqrt{2}\}$ is linearly independent over F_3, we have

$$\begin{array}{ll} & y = 0 \\ \text{and} & x^2 + 1 = 0 . \end{array}$$

Thus there are no solutions at all. The same conclusion holds over F_p, where p is a prime $\equiv 3 \pmod 8$.

These examples should give an indication of why it seems reasonable that we should impose the condition that $Y^d - f(X)$ be absolutely irreducible, i.e. irreducible over F_q and every algebraic extension of F_q, in order to draw the conclusion that the number of solutions be approximately equal to q.

THEOREM 2A. Suppose that $Y^d - f(X)$ is absolutely irreducible and that $q > 100\,dm^2$ where $m = \deg f$. If N is the number of zeros of the polynomial, then

$$|N - q| \le 4\,d^{3/2}\,m\sqrt{q} .$$

Note. No particular importance is attached to the specific values $100\,dm^2$ and $4d^{3/2}m$. This theorem was proved but with different values of the constants) in an elementary way b Stepanov in (1969) for $d = 2$, m odd and q a prime, then in (1970) for $(m,d) = 1$ and q a prime, finally in (1972a) for $d = 2$, m odd and q an arbitrary prime power.

A somewhat sharper estimate will be derived in §11 of Ch. II. The elliptic case of the theorem was first proved by Hasse (1936b,c). The theorem is a special case of Weil's famous theorem (1940, 1948a) on equations $f(x,y) = 0$, which will be proved in Chapter III.

The proof of Theorem 2A will be carried out in the next sections.

LEMMA 2B. Suppose

$$X' = aX + bY + c$$

$$Y' = dX + eY + f$$

is a non-singular linear substitution; i.e., $\begin{vmatrix} a & b \\ d & e \end{vmatrix} \neq 0$, *with coefficients,* a, b, c, d, e, f *in some field* k. *Let* $f(X,Y)$ *be a polynomial with coefficients in* k. *Then* $f(X,Y)$ *is irreducible over* k *if and only if* $f(aX + bY + c,\ dX + eY + f)$ *is irreducible over* k.

Proof: Exercise.

LEMMA 2C. Suppose the polynomial $Y^d - f(X)$ *has coefficients in a field* k. *Then the following three conditions are equivalent:*

(i) $Y^d - f(X)$ *is absolutely irreducible.*

(ii) $Y^d - cf(X)$ *is absolutely irreducible for every* $c \neq 0$, $c \in k$.

(iii) *If* $f(X) = a(X - x_1)^{d_1} \ldots (X - x_s)^{d_s}$ *is the factorization of* f *in* $\bar{k}$, *with* $x_i \neq x_j$ $(i \neq j)$, *then* $(d, d_1, d_2, \ldots, d_s) = 1$.

Proof: Each part of the proof will be by contraposition.

(i) ⇒ (ii). Suppose (ii) is not true. Then $Y^d - cf(X)$ is reducible over $\bar{k}$ for some $c \neq 0$, whence

$$c\left(\left(\frac{Y}{\sqrt[d]{c}}\right)^d - f(X)\right)$$

is reducible over $\bar{k}$. By Lemma 2B, $Y^d - f(X)$ is reducible over $\bar{k}$, contradicting (i).

(ii) ⇒ (iii). Suppose (iii) is not true. Let $t = (d, d_1, \ldots, d_s) > 1$.

Put $$g(X) = (X - x_1)^{d_1/t} \ldots (X - x_s)^{d_s/t} .$$

Then $Y^d - \frac{1}{a} f(X) = Y^d - g(X)^t$

$$= (Y^{d/t} - g(X))(Y^{\frac{d}{t}(t-1)} + Y^{\frac{d}{t}(t-2)} g(X) + \ldots + g(X)^{t-1}).$$

So with $c = \frac{1}{a} \neq 0$, $Y^d - cf(X)$ is reducible in $\bar{k}$, contradicting (ii).

(iii) ⇒ (i). Consider $Y^d - f(X)$ as a polynomial in the ring $L[Y]$, with coefficients in the field $L = \bar{k}(X)$. We then have a factorization over $\bar{L}$:

$$Y^d - f(X) = (Y - \mathfrak{Y}_1) \ldots (Y - \mathfrak{Y}_d) ,$$

where $\mathfrak{Y}_1, \ldots, \mathfrak{Y}_d$ are "algebraic functions"; i.e., elements of $\bar{L}$. In fact, we may set

$$\mathfrak{Y}_1 = \zeta_1 \mathfrak{Y}, \ldots, \mathfrak{Y}_d = \zeta_d \mathfrak{Y} ,$$

where $\mathfrak{Y}$ is any root of $Y^d - f(X)$ in $\bar{L}$, and where $\zeta_1, \ldots, \zeta_d$ are elements of $\bar{k}$ defined by

$$Y^d - 1 = (Y - \zeta_1) \dots (Y - \zeta_d) .$$

Suppose that $Y^d - f(X)$ is reducible over $\bar{k}$. Then there exists a product

$$(Y - \zeta_{i_1}\mathfrak{Y}) \dots (Y - \zeta_{i_h}\mathfrak{Y}) \in \bar{k}[X,Y]$$

where $h < d$. The constant term of this product, $\pm \zeta_{i_1}\zeta_{i_2} \dots \zeta_{i_h}\mathfrak{Y}^h \in \bar{k}[X]$, whence $\mathfrak{Y}^h \in \bar{k}[X]$. Let ℓ be the smallest positive integer with $\mathfrak{Y}^\ell \in \bar{k}[X]$. Then any integer m with $\mathfrak{Y}^m \in \bar{k}[X]$ is a multiple of ℓ . Since $\mathfrak{Y}^d \in \bar{k}[X]$, it follows that $\ell | d$, and since $\mathfrak{Y}^h \in \bar{k}[X]$, $\ell < d$. Say $\mathfrak{Y}^\ell = h(X)$. We have $\mathfrak{Y}^d = f(X)$, so

$$h(X)^{d/\ell} = f(X) .$$

Take $t = \frac{d}{\ell}$; then $t | d_i$ $(i = 1,\dots,s)$, $t > 1$. So $t | (d,d_1,\dots,d_s)$, $t > 1$, and the lemma is established.

COROLLARY. *Suppose* $\deg f = m$. *Then* $Y^d - f(X)$ *is absolutely irreducible if* $(m,d) = 1$.

Note: Rather than the more general condition of absolute irreducibility adopted here, Stepanov always assumed $(m,d) = 1$.

LEMMA 2D. *Let* $\mathbf{C}$ *be a cyclic group of order* h . *For any integer* $d > 0$, *let* $\mathbf{C}^d$ *be the subgroup of* d^{th} *powers of elements of* $\mathbf{C}$. *Let* $d' = (h,d)$. *Then* $\mathbf{C}^d = \mathbf{C}^{d'}$, *and consists precisely of those* $x \in \mathbf{C}$ *with*

(2.1) $$x^{h/d'} = 1 .$$

For any $x \in \underline{\underline{C}}^d$, there are exactly d' elements $y \in \underline{\underline{C}}$ with $y^d = x$.

Proof: Write $\underline{\underline{C}} = \{g, g^2, \ldots, g^h = 1\}$. Suppose $x \in \underline{\underline{C}}^d$, hence is of the form $x = g^{id}$, for some i . Then since $d' | d$,

$$x^{h/d'} = \left(g^{\frac{id}{d'}}\right)^h = 1 .$$

Conversely, suppose $x^{h/d'} = 1$. We must show there is a $y \in \underline{\underline{C}}$ with $y^d = x$. Let $x = g^i$. Then $g^{\frac{ih}{d'}} = 1$; it follows that $\frac{i}{d'}$ is an integer, say, $i = d' i_o$. If $y = g^j$, we need

$$g^{jd} = x = g^{i_o d'} ,$$

or $$jd \equiv i_o d' \pmod h .$$

This congruence has a solution j since $(d,h) = d'$ divides $i_o d'$ Moreover, the number of solutions $j \pmod h$ equals $(d,h) = d'$. Since (2.1) depends only on d', we have $\underline{\underline{C}}^d = \underline{\underline{C}}^{d'}$, and the lemma is proved.

Given an equation $y^d = f(x)$ in F_q , we are interested in the number $N = N(d)$ of solutions (x,y) with components in F_q . Let N_0 be the number of solutions with $y = 0$; then N_0 is the number of $x \in F_q$ with $f(x) = 0$.

Now consider the number of solutions with $y \neq 0$. For such a solution, $f(x) \in (F_q^*)^d$, so by Lemma 2D,

$$f(x)^{\frac{q-1}{d'}} = 1 \, , \quad \text{where} \quad d' = (q-1, d) \, .$$

Let N_1 be the number of $x \in F_q$ with $f(x)^{\frac{q-1}{d'}} = 1$. For such an x, there are d' elements y with $y^d = f(x)$. Hence,

$$N = N_0 + d'N_1 \, .$$

This expression depends only upon d', so $N = N(d')$. Without loss of generality, we may therefore assume that $d \mid (q-1)$; then

$$N = N_0 + dN_1 \, ,$$

where N_1 is the number of x such that $f(x)^{\frac{q-1}{d}} = 1$. Finally, let N_2 be the number of $x \in F_q$ with

$$(2.2) \qquad \left(f(x)^{\frac{q-1}{d}}\right)^{d-1} + \left(f(x)^{\frac{q-1}{d}}\right)^{d-2} + \dots + f(x)^{\frac{q-1}{d}} + 1 = 0 \, .$$

But we have

$$Z^q - Z = Z\left(Z^{\frac{q-1}{d}} - 1\right)\left(Z^{\frac{q-1}{d}(d-1)} + Z^{\frac{q-1}{d}(d-2)} + \dots + Z^{\frac{q-1}{d}} + 1\right) .$$

Now, since every $z \in F_q$ satisfies $z^q - z = 0$, and $Z^q - Z$ is a separable polynomial, every element of F_q is a root of one and only one of the factors of $Z^q - Z$, whence

$$q = N_0 + N_1 + N_2 \, .$$

For future reference, we summarize:

<u>LEMMA 2E</u>: <u>Let</u> N <u>be the number of solutions</u> $(x,y) \in F_q \times F_q$ <u>of</u> $y^d = f(x)$, <u>where</u> $d \mid (q-1)$. <u>Then</u> $N = N_0 + dN_1$, <u>where</u>

N_0 is the number of $x \in F_q$ with $f(x) = 0$, and N_1 is the number of $x \in F_q$ with $f(x)^{\frac{q-1}{d}} = 1$. Further, $N_0 + N_1 + N_2 = q$, where N_2 is the number of x satisfying (2.2).

§ 3. Construction of certain polynomials.

In order to prove Theorem 2A, we may clearly suppose

(3.1) $$m > 1, \quad d > 1 .$$

We assume $d \mid (q-1)$, and, for the moment, that $(d,m) = 1$, where $m = \deg f$. Also assume temporarily that $q = p$ or p^2, p prime. For convenience let

(3.2) $$g(X) = f(X)^{\frac{q-1}{d}} .$$

LEMMA 3A: Suppose $h_0(X), h_1(X), \ldots, h_{d-1}(X)$ are polynomials of the type

$$h_i(X) = k_{i0}(X) + X^q k_{i1}(X) + \ldots + X^{qK} k_{iK}(X)$$

for $0 \le i \le d-1$, and where $\deg k_{ij} \le \frac{q}{d} - m$. If

$$h_0(X) + g(X) h_1(X) + \ldots + g(X)^{d-1} h_{d-1}(X) = 0 ,$$

then each polynomial $k_{ij}(X) = 0$ $(0 \le i \le d-1,\ 0 \le j \le K)$.

Proof: A typical summand is of the form

$$\ell_{ij}(X) = g(X)^i X^{qj} k_{ij}(X) .$$

It suffices to show that the degrees of the non-zero summands are all distinct. We have

$$\deg \ell_{ij} = qj + i\,\frac{q-1}{d}\,m + \deg k_{ij}$$

$$= \frac{q}{d}\,(dj + im) + \deg k_{ij} - \frac{i}{d}\,m\ ,$$

whence

$$\frac{q}{d}\,(dj + im) - m < \deg \ell_{ij} \le \frac{q}{d}\,(dj + im) + \frac{q}{d} - m\ .$$

Hence we need only show that for pairs $(i,j) \neq (i',j')$, we have $dj + im \neq dj' + i'm$.

So suppose $$dj + im = dj' + i'm\ .$$

Then $$im \equiv i'm \pmod d\ ,$$

so since $(m,d) = 1$, $$i \equiv i' \pmod d\ .$$

But $0 \le i,\ i' \le d - 1$, so $i = i'$ and $j = j'$.

LEMMA 3B: (Fundamental lemma). Let ε be an integer, $1 \le \varepsilon \le d - 1$, and let $a(Z)$ be a polynomial of degree ε. Let $\mathfrak{S}$ be the set of $x \in F_q$ with either $a(g(x)) = 0$ or $f(x) = 0$. Let $M \ge m + 1$ be an integer with

$$(M + 3)^2 \le \frac{2q}{d}\ .$$

Then there exists a polynomial $r(X) \neq 0$, which has a zero of order $\ge M$ for every $x \in \mathfrak{S}$ and has

$$\deg r \le \frac{\varepsilon}{d}\,qM + 4mq\ .$$

Proof: Let us try

$$r(X) = f(X)^M \sum_{i=0}^{d-1} \sum_{j=0}^{K} k_{ij}(X)\,g(X)^i X^{qj}\ ,$$

where the $k_{ij}(X)$ are polynomials with coefficients to be determined and $\deg k_{ij} \le \frac{q}{d} - m$, and where

(3.3) $$K = \left[\frac{\varepsilon}{d}(M + m + 1)\right],$$

"[ ]" denoting the integer part. If D is the differentiation operator, then one finds by induction on ℓ for $0 \le \ell \le M - 1$, that

$$D^{\ell} r(X) = f(X)^{M-\ell} \sum_{i=0}^{d-1} \sum_{j=0}^{K}{}' k_{ij}^{(\ell)}(X)\, g(X)^{i} X^{qj},$$

where

$$k_{ij}^{(\ell+1)}(X) = f(X)(D k_{ij}^{(\ell)}(X)) + (D f(X))(M - \ell + i \frac{q-1}{d}) k_{ij}^{(\ell)}(X).$$

Hence $k_{ij}^{(\ell+1)}$ is a polynomial and

$$\deg k_{ij}^{(\ell+1)}(X) \le \deg k_{ij}^{(\ell)}(X) + m - 1 .$$

In particular,

$$\begin{aligned} \deg k_{ij}^{(\ell)}(X) &\le \deg k_{ij}(X) + \ell(m-1) \\ &\le \frac{q}{d} - m + \ell(m-1) \\ &< \frac{q}{d} + \ell(m-1) - 1, \end{aligned}$$

by (3.1).

Now, by hypothesis, we have $(M + 3)^2 \le \frac{2q}{d}$, so $M < \sqrt{q}$, and since we are dealing with the special case where $q = p$ or p^2, we have $M < p$. Theorem 1G is now applicable and for $x \in \mathfrak{S}$, we want that

$$D^{\ell} r(x) = 0 \quad (0 \le \ell \le M - 1).$$

For any $z \in F_q$ satisfying $a(z) = 0$, we have

$$z^{\varepsilon} = c_0 + c_1 z + \dots + c_{\varepsilon-1} z^{\varepsilon-1} ,$$

since $a(Z)$ is of degree ε. Hence for $i \geq 0$,

$$z^i = c_0^{(i)} + c_1^{(i)} z + \dots + c_{\varepsilon-1}^{(i)} z^{\varepsilon-1} .$$

In particular, for $x \in F_q$ satisfying $a(g(x)) = 0$, we have $x^q = x$ and

$$g(x)^i = \sum_{t=0}^{\varepsilon-1} c_t^{(i)} g(x)^t .$$

Then for such an x,

$$D^{\ell} r(x) = f(x)^{M-\ell} \sum_{t=0}^{\varepsilon-1} s_t^{(\ell)}(x) g(x)^t ,$$

where

$$s_t^{(\ell)}(X) = \sum_{i=0}^{d-1} \sum_{j=0}^{K} c_t^{(i)} k_{ij}^{(\ell)}(X) X^j .$$

So certainly $D^{\ell} r(x) = 0$ for $x \in F_q$, $a(g(x)) = 0$, provided the polynomials

$$s_t^{(\ell)}(X) \qquad (0 \leq t \leq \varepsilon - 1)$$

are all identically zero.

Notice that

$$\deg s_t^{(\ell)} < \frac{q}{d} + \ell(m - 1) - 1 + K .$$

Now, if we denote by B the number of coefficients of $s_t^{(\ell)}$

for $0 \leq t \leq \varepsilon - 1$, $0 \leq \ell \leq M - 1$, then

$$B < \varepsilon M\left(\frac{q}{d} + K\right) + \frac{M^2}{2}(m-1)\varepsilon$$

$$< \frac{\varepsilon q}{d} M + \varepsilon M^2\left(\frac{m-1}{2} + \frac{\varepsilon}{d}\right) + \varepsilon M(m+1) .$$

$$< \frac{\varepsilon q}{d} M + \varepsilon M^2 \frac{m+1}{2} + \varepsilon M(m+1)$$

by (3.3).

If we denote by A the number of possible coefficients of all the k_{ij}, then

$$A \geq \left(\frac{q}{d} - m\right) d(K+1)$$

$$\geq (q - md)\frac{\varepsilon}{d}(M + m + 1)$$

$$\geq \frac{\varepsilon q}{d} M + \frac{\varepsilon q}{d}(m+1) - m\varepsilon(2M) ,$$

since $M \geq m + 1$. If it is the case that $B < A$, then the number of conditions on the coefficients of k_{ij} is less than the number of available coefficients of k_{ij}. Since the conditions are homogeneous linear equations, we can then obtain a non-trivial solution for these coefficients. In order that $B < A$, if suffices that

$$M^2\left(\frac{m+1}{2}\right) + 3M(m+1) < \frac{q}{d}(m+1) ,$$

or that

$$M^2 + 6M < \frac{2q}{d} .$$

This is guaranteed by our hypothesis that $(M+3)^2 \leq \frac{2q}{d}$.

We constructed $r(X)$ such that it has a zero of order $\geqq M$ for $x \in F_q$ with $a(g(x)) = 0$. Since $r(X)$ has a factor $f(X)^M$, it is clear that $r(X)$ has a zero of order at least M for each $x \in \mathfrak{S}$. By Lemma 3A, $r(X) \neq 0$.

Finally,

$$\deg r(X) \leq mM + \frac{q}{d} - m + (d-1)m\frac{(q-1)}{d} + qK$$

$$\leq \frac{\varepsilon}{d}qM + q(\frac{1}{d} + m + m + 1) + mM$$

$$\leq \frac{\varepsilon}{d}qM + 4mq ,$$

and the lemma is proved.

§ 4. Proof of the Main Theorem.

In Lemma 3B, the polynomial $r(X)$ was constructed with a zero of order at least M for every $x \in \mathfrak{S}$. But obviously the number of zeros of $r(X)$, counted with multiplicities, cannot exceed its degree; hence,

$$|\mathfrak{S}| \cdot M \leq \deg r \leq \frac{\varepsilon}{d}qM + 4qm ,$$

or

$$|\mathfrak{S}| \leq \frac{\varepsilon}{d}q + 4q\frac{m}{M} .$$

Now choose

$$M = \left[\sqrt{\frac{2q}{d}}\right] - 3 .$$

By the assumption of Theorem 2A that $q > 100\,dm^2$,

$$M \geq \sqrt{\frac{2q}{d}} - 4 \geq \sqrt{\frac{q}{d}} \geq m + 1 .$$

Therefore

$$|\mathfrak{S}| \leq \frac{\varepsilon}{d}q + 4md^{\frac{1}{2}}q^{\frac{1}{2}} .$$

First choose $a(Z) = Z - 1$; here $\varepsilon = 1$. Observe that $\mathfrak{S}$ is the set of $x \in F_q$ with either $g(x) = 1$ or $f(x) = 0$. Thus

$$|\mathfrak{S}| = N_1 + N_0 \leq \frac{q}{d} + 4\,md^{\frac{1}{2}}\,q^{\frac{1}{2}} ,$$

whence

$$(4.1) \qquad N = dN_1 + N_0 \leq d|\mathfrak{S}| \leq q + 4md^{3/2}\,q^{1/2} .$$

Secondly, choose $a(Z) = Z^{d-1} + \ldots + Z + 1$. Here $\varepsilon = d - 1$. Now,

$$\mathfrak{S} = \{x \in F_q :\ g(x)^{d-1} + \ldots + g(x) + 1 = 0 \ \text{ or } \ f(x) = 0\} .$$

Therefore,

$$|\mathfrak{S}| = N_2 + N_0 \leq \frac{d-1}{d}\,q + 4md^{\frac{1}{2}}\,q^{\frac{1}{2}} .$$

But

$$N_1 = q - N_0 - N_2 \geq \frac{q}{d} - 4\,md^{\frac{1}{2}}\,q^{\frac{1}{2}} ,$$

whence

$$(4.2) \qquad N \geq dN_1 \geq q - 4\,md^{3/2}\,q^{1/2} .$$

Finally, combining (4.1) and (4.2),

$$|N - q| \leq 4\,md^{3/2}\,q^{1/2} .$$

This does not, however, complete the proof of Theorem 2A in its generality. It has only been proved under the two assumptions that $(m,d) = 1$ and $q = p$ or p^2. We shall proceed to remove these conditions.

§ 5. Removal of the condition $(m,d) = 1$.

The condition that $(m,d) = 1$ was only required in the

proof of Lemma 3A. The task before us is to prove this lemma under the condition that $Y^d - f(X)$ is absolutely irreducible.

Remark: Recall that $h_i(X)$ was a polynomial of the type

$$h_i(X) = k_{i0}(X) + X^q k_{i1}(X) + \dots + X^{qK} k_{iK}(X) ,$$

where

$$\deg k_{ij} \leq \frac{q}{d} - m .$$

It is easy to see that for $c \in F_q$, $h_i(X - c)$ is a polynomial of the same type. Hence, we may make a substitution $X \to X - c$, and replace the polynomial $f(X)$ by $f(X - c)$. If $q > m$, we may choose $c \in F_q$ such that $f(-c) \neq 0$. Therefore without loss of generality, we assume $f(0) \neq 0$.

First, we consider the case $d = 2$. Assume that $Y^2 - f(X)$ is absolutely irreducible and suppose

$$(5.1) \qquad h_0(X) + h_1(X)\, g(X) = 0 ,$$

or

$$h_0(X) = -h_1(X) f(X)^{\frac{q-1}{2}} .$$

Squaring, we obtain

$$h_0^2(X) f(X) = h_1^2(X) f(X)^q .$$

Then, for some polynomial $\ell(X)$,

$$k_{00}^2(X) f(X) = k_{10}^2(X) f(0)^q + X^q \ell(X) = k_{10}^2(X) f(0) + X^q \ell(X) .$$

Here

$$\deg k_{00}^2(X) f(X) \leqq q - 2m + m = q - m < q ,$$

$$\deg k_{10}^2(X) f(0) \leqq q - 2m < q .$$

It follows that

$$k_{00}^2(X)f(X) = k_{10}^2(X)f(0) \quad .$$

If $k_{00}(X) \neq 0$,

$$f(X) = \left(\sqrt{f(0)} \quad \frac{k_{10}(X)}{k_{00}(X)}\right)^2 ,$$

which is impossible, since $Y^2 - f(X)$ is absolutely irreducible. Therefore, $k_{00}(X) = 0$ and $k_{10}(X) = 0$, since $f(0) \neq 0$. Then dividing (5.1) by X^q and repeating the argument, we conclude that $k_{01} = k_{11} = 0$. Continuing in this way we see that all the k_{ij} are 0 .

For consideration of the general case $d > 2$, we state, without proof, the fundamental theorem on symmetric polynomials.

LEMMA 5A: *Suppose* $a(X_1,\ldots,X_d)$ *is a symmetric polynomial (i.e., invariant under any permutation of the variables) with coefficients in any field. Then there exists a polynomial* $b(U_1,\ldots,U_d)$, *with coefficients in the same field, such that*

$$a(X_1,\ldots,X_d) = b(s_1(X_1,\ldots,X_d),\ldots,s_d(X_1,\ldots,X_d)) ,$$

where

$$s_1 = -(X_1 + X_2 + \ldots + X_d) ,$$

$$s_2 = X_1X_2 + \ldots + X_{d-1}X_d ,$$

$$\vdots$$

$$s_d = (-1)^d X_1X_2 \ldots X_d \quad .$$

Moreover,

(a) *If* $a(X_1,\ldots,X_d)$ *is of degree* δ *in each* X_i , *then* b *is of total degree* δ .

(b) *If* $a(X_1,\dots,X_d)$ *is of total degree* ε *, then each monomial* $U_1^{i_1} \dots U_d^{i_d}$ *of* b *with non-zero coefficients has the property that*

$$i_1 + 2i_2 + \dots + di_d = \varepsilon \ .$$

Form a polynomial

$$a(Y\,;H_0,\dots,H_{d-1}) = H_0 + H_1 Y + \dots + H_{d-1}Y^{d-1} \ .$$

Let $\zeta_1,\dots,\zeta_d$ be elements of $\overline{F_q}$ with

$$X^d - 1 = (X - \zeta_1) \dots (X - \zeta_d) \ ,$$

and put

$$b(Y\,;H_0,\dots,H_{d-1}) = \prod_{i=1}^{d} a(\zeta_i Y\,;H_0,\dots,H_{d-1}) \ .$$

Then b is a polynomial symmetric in $\zeta_1 Y,\dots,\zeta_d Y$. By Lemma 5A, b must be a polynomial in the elementary symmetric functions $s_1,\dots,s_d$ of $\zeta_1 Y,\dots,\zeta_d Y$. But in our case, $s_1 = \dots = s_{d-1} = 0$ and $s_d = -Y^d$, so that

$$b(Y\,;H_0,\dots,H_{d-1}) = c(Y^d\,;H_0,\dots,H_{d-1}) \ .$$

Here $c(W;H_0,\dots,H_{d-1})$ is a polynomial of degree $d - 1$ in W , and of degree d in $H_0,\dots,H_{d-1}$. Now set

$$d(U,V;H_0,\dots,H_{d-1}) = V^{d-1}c(U/V;H_0,\dots,H_{d-1}) \ .$$

Then d is a form of degree $d - 1$ in U, V ; and of degree d in $H_0,\dots,H_{d-1}$.

We now assume $Y^d - f(X)$ to be absolutely irreducible. Suppose

(5.2) $$h_0(X) + h_1(X)g(X) + \dots + h_{d-1}(X)g(X)^{d-1} = 0 .$$

With the above notation,

$$a(g(X) ; h_0(X),\dots,h_{d-1}(X)) = 0 ,$$

and we obtain

$$c(g(X)^d ; h_0(X),\dots,h_{d-1}(X)) = 0 .$$

Recalling that $g(X) = f(X)^{\frac{q-1}{d}}$, we obtain $g(X)^d = f(X)^q/f(X)$ and

$$d(f(X)^q, f(X); h_0(X),\dots,h_{d-1}(X)) = 0 .$$

Collecting all terms with no factor of X^q,

(5.3) $$d(f(0), f(X); k_{00}(X),\dots,k_{d-1,0}(X)) + X^q \ell(X) = 0 ,$$

for some polynomial ℓ. Now,

(5.4) $$d(f(0), f(X); k_{00}(X),\dots,k_{d-1,0}(X))$$

is of degree $d - 1$ in $f(0)$, $f(X)$, and of degree d in $k_{00},\dots,k_{d-1,0}$. But $\deg k_{ij} \le \frac{q}{d} - m$, so the polynomial (5.4) is of degree $\leqq (d-1)m + d\left(\frac{q}{d}-m\right) < q$. Hence by (5.3),

$$d(f(0), f(X); k_{00}(X),\dots,k_{d-1,0}(X)) = 0 .$$

Let $\mathfrak{Y}$ be the algebraic function with

$$\mathfrak{Y}^d = \frac{f(X)}{f(0)} .$$

$\mathfrak{Y}$ is of degree d over $\overline{F}_q(X)$, since $Y^d - \frac{1}{f(0)} f(X)$ is absolutely irreducible. Retracing our steps, we must have

$$c\left(\frac{f(0)}{f(X)}\,;\,k_{00}(X),\dots,k_{d-1,0}(X)\right) = 0 \quad ,$$

or
$$c\left(\frac{1}{\mathfrak{Y}^d}\,;\,k_{00}(X),\dots,k_{d-1,0}(X)\right) = 0 \quad ,$$

and
$$b\left(\frac{1}{\mathfrak{Y}}\,;\,k_{00}(X),\dots,k_{d-1,0}(X)\right) = 0 \quad .$$

Therefore, some factor

$$a\left(\frac{\zeta}{\mathfrak{Y}}\,;\,k_{00}(X),\dots,k_{d-1,0}(X)\right) = 0 \quad ,$$

or
$$k_{00}(X) + \frac{\zeta}{\mathfrak{Y}}\,k_{10}(X) + \dots + \left(\frac{\zeta}{\mathfrak{Y}}\right)^{d-1} k_{d-1,0}(X) = 0 \quad .$$

But $\mathfrak{Y}$ is algebraic of degree d over $\overline{F_q}(X)$, so that

$$k_{00}(X) = \dots = k_{d-1,0}(X) = 0 \ .$$

Now divide (5.2) by X^q and proceed similarly to conclude that

$$k_{01}(X) = \dots = k_{d-1,1}(X) = 0 \ .$$

Continuing in this way we see that all the $k_{ij}(X)$ are zero. We have shown that Lemma 3A holds under the condition that $Y^d - f(X)$ is absolutely irreducible.

§6. Hyperderivatives.

Let k be a field. The polynomial ring $k[X]$ is a vector space over k. Let $E^{(\ell)}$ $(\ell = 0,1,\dots)$ be the linear operator on $k[X]$ with

$$E^{(\ell)}(X^t) = \binom{t}{\ell} X^{t-\ell} \qquad (t = 0,1,\dots) \quad .$$

If D is the differentiation operator, then $D^{\ell}(X^t) = \ell!\binom{t}{\ell}X^{t-\ell}$,

and hence $D^{\ell} = \ell!\,E^{(\ell)}$. Thus if k is of characteristic 0, then

$$E^{(\ell)} = \frac{1}{\ell!}D^{\ell} .$$

We call the operators $E^{(\ell)}$ hyperderivatives. They are also called Hasse derivatives. See the papers Hasse (1936a), Teichmüller (1936).

LEMMA 6A.

$$E^{(\ell)}(f_1(X)\ldots f_t(X)) = \sum_{\substack{i_1\geqq 0,\ldots,i_t\geqq 0\\ i_1+\ldots+i_t=\ell}} E^{(i_1)}(f_1(X))\ldots E^{(i_t)}(f_t(X)) .$$

Proof. It will suffice to prove the case $t = 2$, since the general case follows by an obvious induction on t. Thus we have to show that

$$E^{(\ell)}(f(X)g(X)) = \sum_{i=0}^{\ell} E^{(i)}(f(X))E^{(\ell-i)}(g(X)) . \tag{6.1}$$

By the linearity of $E^{(j)}$, we may suppose that $f(X)$, $g(X)$ are monomials; say $f(X) = X^a$, $g(X) = X^b$. Then (6.1) is equivalent to

$$\binom{a+b}{\ell} = \sum_{i=0}^{\ell}\binom{a}{i}\binom{b}{\ell-i} .$$

But this identity is an immediate consequence of the definition of $\binom{a+b}{\ell}$ as the number of subsets with ℓ elements contained in a set of $a + b$ elements.

COROLLARY 6B. $E^{(\ell)}(X - c)^t = \binom{t}{\ell}(X - c)^{t-\ell}$.

Proof. $$E^{(\ell)}(X-c)^t = \sum_{\substack{i_1 \geqq 0, \ldots, i_t \geqq 0 \\ i_1 + \ldots + i_t = \ell}} (E^{(i_1)}(X-c)) \ldots (E^{(i_t)}(X-c)).$$

Now $E^{(1)}(X - c) = 1$ and $E^{(i)}(X - c) = 0$ if $i \geqq 2$. Hence in the above sum, we need only consider summands with each i_j either 0 or 1 . The number of such summands is $\binom{t}{\ell}$, and each summand is $(X - c)^{t-\ell}$.

COROLLARY 6C. *Suppose* $0 \leqq \ell \leqq t$. *Then*

$$E^{(\ell)}(a(X)f(X)^t) = b(X)f(X)^{t-\ell}, \tag{6.2}$$

where $b(X)$ *is a polynomial with*

$$\deg b = \deg a + \ell((\deg f) - 1) .$$

Proof. In

$$E^{(\ell)}(a(X)f(X)^t) = \sum_{\substack{i_0 \geqq 0, \ldots, i_t \geqq 0 \\ i_0 + \ldots + i_t = \ell}} (E^{(i_0)}a(X))(E^{(i_1)}f(X)) \ldots (E^{(i_t)}(f(X)) ,$$

every summand is divisible by $f(X)^{t-\ell}$. Hence a formula such as (6.2) holds. Furthermore,

$$\begin{aligned}\deg b &= \deg(E^{(\ell)}(af^t)) - (t-\ell)\deg f \\ &= \deg a + t\deg f - \ell - (t-\ell)\deg f \\ &= \deg a + \ell(\deg f - 1)\ .\end{aligned}$$

THEOREM 6D. *Suppose* $E^{(\ell)}(f(x)) = 0$ *for* $\ell = 0,1,\dots,M-1$. *Then* $(X-x)^M$ *divides* $f(X)$.

Proof. We may write $f(X) = a_0 + a_1(X-x) + \dots + a_d(X-x)^d$. By Corollary 6B,

$$E^{(\ell)}f(X) = a_\ell + \binom{\ell+1}{\ell}a_{\ell+1}(X-x) + \dots + \binom{d}{\ell}a_d(X-x)^{d-\ell}\ .$$

The hypothesis of the lemma implies that $a_\ell = 0$ for $\ell = 0,1,\dots,M-1$, and the conclusion follows.

LEMMA 6E. *Suppose* k *is of characteristic* $p > 0$. *Let*

$$r(X) = h(X, X^{p^\mu})$$

for some polynomial $h(X,Y)$. *Then for* $\ell < p^\mu$,

$$E^{(\ell)}r(X) = E_X^{(\ell)}h(X, X^{p^\mu})\ ,$$

where $E_X^{(\ell)}$ *is the "partial" hyperderivative with respect to* X *of* $h(X,Y)$.

Proof. By linearity, it suffices to take the case when $h(X,Y) = X^aY^b$. Then by Lemma 6A it suffices to show that for $0 < \ell < p^\mu$,

$$E^{(\ell)}(X^{p^\mu}) = 0\ .$$

This in turn follows from the fact that $\binom{p^\mu}{\ell} = (p^\mu/\ell)\binom{p^\mu - 1}{\ell - 1}$ is 0 in a field of characteristic p.

§7. Removal of the condition that $q = p$ or p^2.

We just have to prove Lemma 3B in general. We set up

$$r(X) = h(X, X^q)$$

with

$$h(X,Y) = f(X)^M \sum_{i=0}^{d-1} \sum_{j=0}^{K} k_{ij}(X) g(X)^i Y^j .$$

We now simply have to use Theorem 6D instead of Theorem 1G, hence have to compute $E^{(\ell)} r(X)$ instead of $D^\ell r(X)$. By Corollary 6C, and since $g(X)$ is a power of $f(X)$,

$$E^{(\ell)}(f(X)^M k_{ij}(X) g(X)^i) = f(X)^{M-\ell} k_{ij}^{(\ell)}(X) g(X)^i ,$$

where

$$\deg k_{ij}^{(\ell)} \leqq \deg k_{ij} + \ell(m-1) .$$

In view of Lemma 6E we have, for $0 \leqq \ell < M \leqq q = p^\kappa$,

$$E^{(\ell)} r(X) = f(X)^{M-\ell} \sum_{i=0}^{d-1} \sum_{j=0}^{K} k_{ij}^{(\ell)}(X) g(X)^i X^{qj} .$$

The rest of the argument is as in §3 .

§8. The Work of Stark.

Now suppose $d = 2$ and consider again the hyperelliptic equation

$$y^2 = f(x) ,$$

where $f(X)$ is a polynomial of degree m, and where $Y^2 - f(X)$ is absolutely irreducible. We proved that the number N of solutions satisfies

$$|N - q| < 4md^{3/2} q^{1/2} ,$$

if $q > dm^2$.

H. M. Stark (1973) obtained the sharper bounds

$$|N - q| \leqq (m-1)q^{1/2} ,$$

if $q = p$ and if $f(X)$ has m distinct roots. Set

$$g = \begin{cases} \dfrac{m-1}{2} & \text{if } m \text{ is odd,} \\[2ex] \dfrac{m-2}{2} & \text{if } m \text{ is even .} \end{cases}$$

The number g is called the "genus" of the equation. Thus Stark obtains

$$\begin{aligned} |N - q| &\leqq 2g\,q^{1/2} , && \text{if } m \text{ is odd ,} \\ &\leqq (2g+1)q^{1/2} , && \text{if } m \text{ is even .} \end{aligned} \tag{8.1}$$

In fact it follows from Weil's theorem that $|N - q| \leqq 2g\,q^{\frac{1}{2}}$ if m is odd, and $|N - q + 1| \leqq 2g\,q^{\frac{1}{2}}$ if m is even. Moreover, the constant $2g$ cannot be replaced by a smaller constant independent of q.

However, Stark in his paper did in some cases improve on (8.1) if m is odd. For example, he showed that if $m = 5$ (so that $g = 2$) and if q is a prime p of the type $p = 4r^2 + 1$ $(r \geqq 2)$, then

$$|N - p| \leqq 2g[\sqrt{p}] - 1 \quad .$$

He achieved this improvement by permitting polynomials $k_{ij}(X)$ in Lemma 3B whose degree is larger than $\frac{q}{d} - m$. In fact their degrees may exceed $\frac{q}{d}$. But then it is much more difficult to prove that the polynomial $r(X)$ of Lemma 3B is not 0 .

§9. Equations $y^q - y = f(x)$.

The first elementary treatment of such equations is due to Stepanov (1971), with a less complicated treatment provided by Mitkin (1972).

THEOREM 9A: Suppose $r = q^{\kappa}$. Let $f(X) \in F_q[X]$, with $(q, \deg f) = 1$ and $\deg f < q$. If N is the number of solutions $(x,y) \in F_r^2$ of $y^q - y = f(x)$, then

$$|N - r| < q^{[\frac{\kappa}{2}]+4} .$$

Note: This inequality is only significant when κ is large. For example, the theorem yields no information when $\kappa = 2$: we get

$$|N - q^2| < q^5 ,$$

but obviously

$$0 \le N \le |F_r^2| = q^4 .$$

Recall that if $x \in F_r$, then the trace

$$\mathfrak{T}(f(x)) = f(x) + f(x)^q + \ldots + f(x)^{q^{\kappa-1}} \in F_q .$$

For $w \in F_q$, let N_w be the number of $x \in F_r$ with

$$\mathfrak{T}(f(x)) = w .$$

LEMMA 9B.

$$\sum_{w \in F_q} N_w = r \quad \text{and} \quad N = qN_0 .$$

Proof: The first statement is obvious. The fact that $N = qN_0$ follows from Lemma 1F.

Now let $\nu = [\frac{\kappa}{2}]$. We may assume $\kappa \geq 3$; hence $\nu \geq 1$. Let

$$g(X) = f(X)^{q^{\nu}} + f(X)^{q^{\nu+1}} + \dots + f(X)^{q^{\kappa-1}},$$

$$h(X) = f(X) + f(X)^{q} + \dots + f(X)^{q^{\nu-1}}.$$

<u>LEMMA 9C: Let</u> $w \in F_q$ <u>be fixed. Let</u> M <u>be divisible by</u> q, <u>and</u> $0 < M \leq q^{\kappa-\nu-1}$. <u>Then there is a polynomial</u> $u(X) \neq 0$, <u>which has a zero of order</u> $\geq M$ <u>for every</u> $x \in F_r$ <u>with</u>

$$\mathfrak{T}(f(x)) = w,$$

<u>and</u> $\deg u(X) \leq M\frac{r}{q} + q^{\kappa+1}$.

<u>Proof</u>: We try

$$u(X) = \sum_{i=0}^{q-1} \sum_{j=0}^{K} k_{ij}(X)\, g(X)^{i}\, X^{rj},$$

where $K = \frac{M}{q}$, and the polynomials $k_{ij}(X)$ have $\deg k_{ij} < \frac{r}{q} = q^{\kappa-1}$, and coefficients to be determined. Since $\kappa \leq 2\nu + 1$, $M \leq q^{\kappa-\nu-1} \leq q^{\nu}$. Thus for $\ell < M \leq q^{\nu}$ and $\mathcal{U}(X) = a(X, X^{q^{\nu}})$, Lemma 6E (with $\mu = \nu\sigma$ if $q = p^{\sigma}$) yields

$$E^{(\ell)}u(X) = E_X^{(\ell)}a(X, X^{q^{\nu}}).$$

Therefore, since $X^r = X^{q^{\kappa}}$ and since

$$g(X) = f(X^{q^{\nu}}) + \dots + f(X^{q^{\kappa-1}}),$$

it follows that

$$E^{(\ell)}u(X) = \sum_{i=0}^{q-1} \sum_{j=0}^{K} k_{ij}^{(\ell)}(X)\, g(X)^{i} X^{rj}$$

with $k_{ij}^{(\ell)}(X) = E^{(\ell)}k_{ij}(X)$.

We proceed just as in the proof of Lemma 3B. Let A be the total number of available coefficients of the polynomials $k_{ij}(X)$. Then

$$A = q^{K-1}q(K+1) = q^{K-1}M + q^{K} .$$

For $x \in F_r$ with $\mathfrak{T}(f(x)) = w$, we have $x^r = x$ and $w = h(x) + g(x)$. So, $E^{(\ell)}u(x) = s^{(\ell)}(x)$, where

$$s^{(\ell)}(X) = \sum_{i=0}^{q-1} \sum_{j=0}^{K} k_{ij}^{(\ell)}(X)(w - h(X))^i X^j .$$

In view of Theorem 6D, in order that $u(X)$ has a zero of order M for our elements $x \in F_r$ with $\mathfrak{T}(f(x)) = w$, it is certainly sufficient that the polynomials $s^{(\ell)}(X)$ vanish identically. Since $K \le q^{\nu-1}$,

$$\deg s^{(\ell)}(X) \le q^{K-1} + (q-1)^2 q^{\nu-1} + K$$
$$\le q^{K-1} + q^{\nu+1} - 2 .$$

Let B denote the total number of conditions (clearly in the form of linear homogeneous equations) on the coefficients of the k_{ij}. If, for each ℓ, $0 \le \ell \le M-1$, we try to make $s^{(\ell)}(X) = 0$, then the number of conditions for this fixed ℓ is at most $\deg s^{(\ell)}(X) + 1 \leqq q^{K-1} + q^{\nu+1} - 1$. Hence

$$B < M(q^{K-1} + q^{\nu+1}) \leqq Mq^{K-1} + q^{K-\nu-1}q^{\nu+1} = Mq^{K-1} + q^{K} .$$

Thus $B < A$, and we may choose the coefficients of $k_{ij}(X)$, not all zero, so that $u(X)$ as a zero of order at least M for the elements x in question. Moreover,

$$\deg u(X) \le rK + (q-1)^2 q^{K-1} + q^{K-1}$$
$$\le M\frac{r}{q} + q^{K+1} .$$

Finally, $u(X)$ does not vanish identically, because the non-zero summands

$$\ell_{ij}(X) = k_{ij}(X)\, g(X)^{i} X^{rj}$$

have degrees

$$\begin{aligned}\deg \ell_{ij}(X) &= rj + iq^{\kappa-1} \deg f + \deg k_{ij} \\ &= q^{\kappa-1}(qj + i \deg f) + \deg k_{ij},\end{aligned}$$

which are distinct by the same argument as in Lemma 3A. We only have to observe that q and $\deg f$ are coprime.

Proof of Theorem 9A: For fixed $w \in F_q$,

$$N_w \cdot M \le \deg r \le M \frac{r}{q} + q^{\kappa+1},$$

or

$$N_w \le \frac{r}{q} + \frac{q^{\kappa+1}}{M}.$$

Choose $M = q^{\kappa-\nu-1}$; then for $\kappa \ge 3$, $q \mid M$. We obtain

$$N_w \le \frac{r}{q} + q^{\nu+2},$$

and by Lemma 9B,

$$N_w = r - \sum_{v \ne w} N_v > \frac{r}{q} - q^{\nu+3}.$$

So

$$\left| N_w - \frac{r}{q} \right| < q^{\nu+3},$$

and, in particular,

$$\left| N_0 - \frac{r}{q} \right| < q^{\nu+3}.$$

By Lemma 9B again,

$$|N - r| < q^{\nu+4} = q^{[\frac{\kappa}{2}]+4}.$$

II. Character Sums and Exponential Sums.

Literature: Weil (1948b), Carlitz and Uchiyama (1957), Perelmuter (1963), Postnikov (1967), Carlitz (1969).

§1. Characters of Finite Abelian Groups.

We now interrupt our investigation of equations over finite fields to deal with character sums and exponential sums. These sums have many applications in analytic number theory.

Given an abelian (multiplicative) group G, a character on G is a map χ from G to the complex numbers with $|\chi(x)| = 1$ for all x and with

$$\chi(xy) = \chi(x)\chi(y)$$

for $x,y \in G$. Since $\chi(1) = \chi(1)\chi(1)$, we have $\chi(1) = 1$.

If χ_1, χ_2 are characters on G, then so is the map $\chi_1\chi_2$ defined by $(\chi_1\chi_2)(x) = \chi_1(x)\chi_2(x)$. If χ is a character, then so is the map χ^{-1} defined by $\chi^{-1}(x) = 1/\chi(x) = \overline{\chi(x)}$ (i.e., the complex conjugate of $\chi(x)$). It is now clear that the characters on G form a group G' under multiplication, whose identity element is the character χ_0 having $\chi_0(x) = 1$ for $x \in G$. The group G' is called the dual group to G.

Write

$$e(x) = e^{2\pi i x} .$$

LEMMA 1A. Let $\underline{C}_n$ be the cyclic group of order n, and let g be a fixed generator. Given a residue class a (modulo n), the map χ_a with

(1.1) $$\chi_a(g^t) = e(at/n) \qquad (t = 0, \pm 1, \ldots)$$

is a character of $\underline{C}_n$. *Every character of* $\underline{C}_n$ *is of this type. The dual group to* $\underline{C}_n$ *is again cyclic of order* n .

Proof. It is readily verified that χ_a , as given by (1.1), is well defined and is a character. It clearly depends only on the residue class of a (modulo n) . For distinct residue classes, one gets distinct characters χ_a . Since $\chi_a \chi_b = \chi_{a+b}$, the characters χ_a form a group which is isomorphic to the integers modulo n , and hence it is cyclic of order n . It remains to be shown that every character χ is a χ_a for some a . Now $\chi(g)^n = \chi(g^n) = \chi(1) = 1$, so that $\chi(g)$ is an nth root of unity, or $\chi(g) = e(a/n)$ for some a . But then $\chi(g^t) = e(at/n)$, and $\chi = \chi_a$.

LEMMA 1B. *Let* $G = G_1 \otimes G_2$ *be the direct product of the abelian groups* G_1 , G_2 . *Then the dual groups* G', G_1' , G_2' *satisfy*

$$G' \cong G_1' \otimes G_2' .$$

Proof. G consists of pairs (x_1, x_2) with $x_1 \in G_1$, $x_2 \in G_2$. With every $\chi_1 \in G_1'$ and $\chi_2 \in G_2'$ we associate the map $\chi: G \to \mathbb{C}$ with $\chi(x_1, x_2) = \chi_1(x_1)\chi_2(x_2)$. It is easily seen that χ is a character of G , and in fact that the map

$$(\chi_1, \chi_2) \to \chi$$

is an isomorphism of $G_1' \otimes G_2'$ into G' . In fact, it is an isomorphism onto, for if $\chi \in G'$, then

$$\chi(x_1, x_2) = \chi(x_1, 1)\chi(1, x_2) = \chi_1(x_1)\chi_2(x_2)$$

with $\chi_1(x_1) = \chi(x_1, 1)$ and $\chi_2(x_2) = \chi(1, x_2)$; clearly $\chi_1 \in G_1'$ and $\chi_2 \in G_2'$.

<u>THEOREM 1C. Given a finite abelian group</u> G , <u>its group</u> G' <u>of characters is isomorphic to</u> G .

<u>Proof.</u> It is well known that every finite abelian group G is of the type $G = \underline{\underline{C}}_{n_1} \otimes \underline{\underline{C}}_{n_2} \otimes \ldots \otimes \underline{\underline{C}}_{n_k}$ for cyclic groups $\underline{\underline{C}}_{n_1}, \ldots, \underline{\underline{C}}_{n_k}$. The theorem now follows from Lemma 1A and repeated application of Lemma 1B.

<u>THEOREM 1D. Let</u> G <u>be a finite abelian group of order</u> $|G|$.

(a) <u>Given a character</u> χ ,

$$\sum_{x \in G} \chi(x) = \begin{cases} |G| & \text{if } \chi = \chi_0 \\ 0 & \text{if } \chi \neq \chi_0 \end{cases} .$$

(b) <u>Given an</u> $x \in G$,

$$\sum_{\chi \in G'} \chi(x) = \begin{cases} |G| & \text{if } x = 1 \\ 0 & \text{if } x \neq 1 . \end{cases}$$

<u>Proof.</u> The assertion (a) is obvious if $\chi = \chi_0$. If $\chi \neq \chi_0$, there exists an $x_1 \in G$ with $\chi(x_1) \neq 1$. As x runs through G , so does xx_1 ; therefore

$$S = \sum_{x \in G} \chi(x) = \sum_{x \in G} \chi(xx_1) = \chi(x_1) S .$$

The desired conclusion $S = 0$ follows from $\chi(x_1) \neq 1$.

Part (b) may be proved in an entirely analogous manner. Or, one may observe that for given x , the map $\chi \to \chi(x)$ is a map from G' into the complex numbers, which is in fact a character on G'. In conjunction with Theorem 1C, one sees that every character of G' is obtained in this way, and that G is therefore the group of characters of G'. The relation between G, G' is thus completely symmetric. Hence (b) follows from (a) if we interchange the roles of G, G'.

§2. Characters and Character Sums associated with Finite Fields.

The non-zero elements of the finite field F_q form a cyclic group F_q^* of $q-1$ elements. Hence the characters χ of F_q^* also form a cyclic group of $q-1$ elements. Thus every character χ will have $\chi^{q-1} = \chi_0$, where χ_0 is the character with $\chi_0(x) = 1$ for all x. We call χ_0 the <u>principal character</u>. We say that χ <u>is of order</u> d if $\chi^d = \chi_0$, and if d is the smallest positive integer with this property. It is easily seen that $d \mid q-1$. We say that χ is <u>of exponent</u> e if $\chi^e = \chi_0$; clearly this is equivalent to $d \mid e$, where d is the order of χ.

Suppose $d \mid q-1$. For every χ of exponent d and every $x \in F_q^*$, we have $\chi(x^d) = \chi(x)^d = \chi^d(x) = 1$. Thus $\chi(y) = 1$ if $y \in (F_q^*)^d$, the group of non-zero d^{th} powers. Conversely, if $\chi(y) = 1$ for every $y \in (F_q^*)^d$, then $\chi^d = \chi_0$. Thus if χ is a character of exponent d, then $\chi(x)$ depends only on the coset of x modulo the subgroup $(F_q^*)^d$. Thus a character of exponent d may be interpreted as a character on the factor group $F_q^*/(F_q^*)^d$. There are precisely d such characters.

It will be convenient to extend the definition of characters χ on F_q^* by putting

$$\chi(0) = \begin{cases} 1 & \text{if } \chi = \chi_0, \\ 0 & \text{if } \chi \neq \chi_0 \end{cases} .$$

We still write $\chi = \chi_1\chi_2$ if $\chi(x) = \chi_1(x)\chi_2(x)$ for $x \in F_q^*$, but not necessarily for $x = 0$. For instance, $\chi^{q-1} = \chi_0$, although $\chi(0) = 0$ for $\chi \neq \chi_0$ and $\chi_0(0) = 1$.

LEMMA 2A. *Suppose* $d \mid q - 1$. *Then*

$$\sum_{\chi \text{ of exponent } d} \chi(x) = \begin{cases} d & \underline{\text{if}} \quad x \in (F_q^*)^d , \\ 0 & \underline{\text{if}} \quad x \notin (F_q^*)^d , \quad x \neq 0 , \\ 1 & \underline{\text{if}} \quad x = 0 . \end{cases}$$

Proof. The characters of exponent d are characters of $F_q^*/(F_q^*)^d$. Hence the first two cases of the lemma follow from Theorem 1D. If $x = 0$, then $\sum_{\chi} \chi(x) = \chi_0(0) + \sum_{\chi \neq \chi_0} \chi(0) = 1 + 0 = 1$.

The characters χ studied so far will henceforth be called the *multiplicative characters* of F_q.

In §3 we shall take the "low road", and we shall easily prove

THEOREM 2B. *Suppose* $d \mid q - 1$ *and suppose* $\chi \neq \chi_0$ *is a character of exponent* d. *Suppose* $f(X)$ *is a polynomial of degree* m *with coefficients in* F_q *and with* $Y^d - f(X)$ *absolutely irreducible. Then if* $q > 100\, dm^2$, *we have*

$$\left| \sum_{x \in F_q} \chi(f(x)) \right| < 5m\, d^{3/2} q^{1/2} . \tag{2.1}$$

This result will turn out to be a consequence of Theorem 2A of Ch. I. We shall also prove

THEOREM 2B'. Suppose χ is a character of order $d > 1$. Suppose $f(X) \in F_q[X]$ is of degree m and is not a dth power, i.e. not of the type $f(X) = c(\ell(X))^d$ with $c \in F_q$ and $\ell(X) \in F_q[X]$. Then if $q > 100\, d\, m^2$, we have again (2.1).

Later on we shall take the "high road" and prove the following sharper results.

THEOREM 2C. Suppose $\chi \neq \chi_\circ$ is a multiplicative character of exponent d. Suppose $f(X) \in F_q[X]$ has precisely m distinct ones among its zeros, and suppose that $Y^d - f(X)$ is absolutely irreducible. Then

$$(2.2) \qquad \Big| \sum_{x \in F_q} \chi(f(x)) \Big| \leq (m-1)q^{1/2} .$$

THEOREM 2C'. Let χ be of order $d > 1$. Suppose $f(X)$ has m distinct ones among its zeros, and it is not a dth power. Then again (2.2) holds.

We now turn to additive characters of F_q. Such an additive character is simply a character of the additive group of F_q. If $q = p^\nu$ where p is the characteristic, then this additive group is the direct sum of ν copies of $\underline{\underline{C}}_p$. Write $\mathfrak{T}$ for the trace from F_q to F_p.

LEMMA 2D. For every $a \in F_q$, the function ψ_a with

$$\psi_a(x) = e(\mathfrak{T}(ax)/p)$$

is an additive character of F_q. Every additive character of F_q is of this type.

<u>Proof</u>. We have $\mathfrak{T}(a(x_1+x_2)) = \mathfrak{T}(ax_1) + \mathfrak{T}(ax_2)$, whence

$$\psi_a(x_1+x_2) = \psi_a(x_1)\psi_a(x_2) .$$

Thus ψ_a is an additive character. By Theorem 1C, the number of additive characters is q ; but so is the number of elements $a \in F_q$. Since, as is easily seen, $\psi_a \neq \psi_{a'}$, if $a \neq a'$, it follows that as a runs through F_q , then ψ_a runs through all additive characters.

Additive characters will always be denoted by the letter ψ . The character ψ_0 with $\psi_0(x) = 1$ for all x is the identity element of the group of additive characters.

THEOREM 2E. *Suppose* $\psi \neq \psi_0$ *is an additive character. Let* $g(X)$ *be a polynomial in* $F_q[X]$ *of degree* n . *Suppose that either*

(i) $n < q$ *and* g.c.d. $(n,q) = 1$, *or, more generally, that*

(ii) $Z^q - Z - g(X)$ *is absolutely irreducible.*

Then *

$$\left| \sum_{x \in F_q} \psi(g(x)) \right| \leq (n-1)q^{1/2} .$$

It will be proved in Theorem 1B of Ch. III that hypothesis (i) implies hypothesis (ii). Strictly speaking, only the case (i) will be proved in this chapter. It will follow from Theorem 9A of Ch. I. The case (ii) depends on results which will be proved in Ch. III. The case (i) is used most often in analytic number theory. In view of Lemma 2D, the case $q=p$ may be reformulated as follows.

COROLLARY 2F. Suppose p is a prime. Suppose $g(X) = a_n X^n + \ldots + a_0$ is a polynomial with integer coefficients having $0 < n < p$ and $p \nmid a_n$. Then

$$\left| \sum_{x=0}^{p-1} e(g(x)/p) \right| \leq (n-1)p^{1/2} .$$

Next, we study "hybrid sums" involving a multiplicative character χ and additive character ψ.

THEOREM 2G. Let χ, ψ be, respectively, a multiplicative character $\neq \chi_0$ of order d with $d \mid q-1$, and an additive character $\neq \psi_0$, of F_q. Let $f(X) \in F_q[X]$ have precisely m distinct ones among this roots, and let $g(X) \in F_q[X]$ have degree n. Suppose that either

(i) $(d, \deg f) = (n,q) = 1$, or, more generally, that

(ii) the polynomials $Y^d - f(X)$ and $Z^q - Z - g(X)$ are absolutely irreducible.

Then

$$\left| \sum_{x \in F_q} \chi(f(x))\psi(g(x)) \right| \leq (m+n-1)q^{1/2} .$$

Again, strictly speaking, the proof of the theorem in this chapter will be not quite complete. We shall need certain results proved only in Ch. VI. It will follow from Theorem 1B in Ch. III that hypothesis (i) implies (ii).

The polynomials $f(X), g(X)$ of our theorems may sometimes be replaced by rational functions. (Perelmuter (1963)) Here we will prove only the following result of this kind.

THEOREM 2H. *Suppose* $\psi \neq \psi_0$ *is an additive character of* F_q. *Suppose* $a, b \in F_q$ *are not both zero. Then*

$$(2.3) \qquad \left| \sum_{x \in F_q^*} \psi(ax + bx^{-1}) \right| \leq 2q^{1/2} .$$

Sums of the type of this theorem are called Kloosterman sums.

All the results enunciated in this section are due to A. Weil (1948b). The proofs of the authors listed at the beginning all follow more or less the same method, but they are given in a more elementary style. In particular, the reference to class field theory is avoided. We shall also present this same method.

Very easy special cases will be given in §3. In §4 we will follow the "low road" to prove Theorems 2B, 2B′. In §5 we will give an application of Theorem 2B′. Finally, in §6-12, we shall deal with the main theorems. In §13 we shall show that Theorem 2E is in a sense best possible.†)

§3. Gaussian Sums.

Before embarking on the more complicated proofs of the theorems announced in the last section, we now pause to prove results of a very simple nature.

The simplest of the hybrid sums introduced in the last section are when $f(X) = g(X) = X$. They are thus of the type

$$G(\chi, \psi) = \sum_{x \in F_q} \chi(x)\psi(x) ,$$

where χ, ψ are a multiplicative and an additive character. These sums are called *Gaussian sums*. In view of Theorem 1D, it is clear that

†) We shall not treat exponential sums along curves (Bombieri (1966) or Chalk and Smith(1971)) or multiple exponential sums (Bombieri(1966) and Deligne(1973)).

(3.1) $$G(\chi_o,\psi) = 0 \quad \text{if} \quad \psi \neq \psi_o ,$$

(3.2) $$G(\chi,\psi_o) = 0 \quad \text{if} \quad \chi \neq \chi_o ,$$

(3.3) $$G(\chi_o,\psi_o) = q .$$

<u>THEOREM 3A.</u> <u>If</u> $\chi \neq \chi_o$ <u>and</u> $\psi \neq \psi_o$, <u>then</u>

$$|G(\chi,\psi)| = q^{1/2} ,$$

Compare with the case $m = n = 1$ of Theorem 2G!

<u>Proof.</u>

$$|G(\chi,\psi)|^2 = \sum_x \sum_y \chi(x)\psi(x)\,\overline{\chi(y)}\;\overline{\psi(y)} .$$

Since $\chi(0) = 0$, we may restrict ourselves to summands with $y \neq 0$
Then $\overline{\chi(y)} = (\chi(y))^{-1} = \chi(1/y)$ and $\overline{\psi(y)} = (\psi(y))^{-1} = \psi(-y)$.
Putting $x = ty$, we obtain

$$|G(\chi,\psi)|^2 = \sum_{y \neq 0} \sum_t \chi(ty)\psi(ty)\chi(1/y)\psi(-y)$$

$$= \sum_t \chi(t) \sum_{y \neq 0} \psi((t-1)y) .$$

$$= \sum_t \chi(t) \sum_y \psi((t-1)y) - (\sum_t \chi(t))$$

$$= \sum_t \chi(t) \sum_y \psi((t-1)y) ,$$

by Theorem 1D. Again by Theorem 1D, the inner sum here is q if $t = 1$, and it is 0 if $t \neq 1$. Thus

$$|G(\chi,\psi)|^2 = \chi(1)q = q .$$

LEMMA 3B. *Suppose* $\psi \neq \psi_0$ *is an additive character. Suppose* $d \mid q-1$, *and suppose* $a \neq 0$ *lies in* F_q . *Then*

$$\sum_{y \in F_q} \psi(ay^d) = \sum_{\chi \text{ of exponent } d} \bar{\chi}(a) G(\chi,\psi) .$$

Proof. For given $x \in F_q$, the number of $y \in F_q$ with $y^d = x$ equals d if $x \in (F_q^*)^d$, it equals 0 if $x \notin (F_q^*)^d$, $x \neq 0$, and it is 1 if $x = 0$. Hence by Lemma 2A,

$$\sum_y \psi(ay^d) = \sum_x \psi(ax) \sum_{\chi \text{ of exp. } d} \chi(x) .$$

Replacing x by x/a and noting that $\chi(x/a) = \chi(x)\bar{\chi}(a)$, we get

$$\sum_x \psi(x) \sum_{\chi \text{ of exp. } d} \chi(x)\bar{\chi}(a)$$

$$= \sum_{\chi \text{ of exp. } d} \bar{\chi}(a) \sum_x \chi(x)\psi(x)$$

$$= \sum_{\chi \text{ of exp. } d} \bar{\chi}(a)\, G(\chi,\psi) .$$

THEOREM 3C. *Suppose* q *is odd,* $\psi \neq \psi_0$ *is an additive character, and* $a \neq 0$, b , c *lie in* F_q . *Then*

$$\left| \sum_{x \in F_q} \psi(ax^2 + bx + c) \right| = q^{1/2} .$$

Compare with the case $n = 2$ of Theorem 2E!

Proof.

$$\sum_x \psi(ax^2 + bx + c) = \sum_x \psi(a(x + \frac{b}{2a})^2 + c - \frac{b^2}{4a})$$

(3.4)

$$= \psi(c - (b^2/4a)) \sum_y \psi(ay^2) .$$

By the case $d = 2$ of Lemma 3B $(2 | q - 1$ since q is odd), we have

$$\sum_y \psi(ay^2) = \sum_{\chi \text{ of exp. } 2} \bar{\chi}(a)\ G(\chi,\psi) . \tag{3.5}$$

There are two characters χ of exponent 2 . One of them is χ_o ; then $G(\chi,\psi) = G(\chi_o,\psi) = 0$ by (3.1) . The other is $\neq \chi_o$; then $|G(\chi,\psi)| = q^{1/2}$ by Theorem 3A. Thus the sum (3.5) is $q^{1/2}$ in absolute value, and the theorem is an immediate consequence of (3.4).

THEOREM 3D. For an additive character $\psi \neq \psi_o$, $a \neq 0$ in F_q and for $d \geq 1$,

$$\left| \sum_{x \in F_q} \psi(ax^d) \right| \leq (d-1) q^{1/2} .$$

Our theorem is a special case of Theorem 2E.

Proof.

$$\sum_x \psi(ax^d) = \sum_x \psi(ax^{d'}) ,$$

where $d' = \text{g.c.d.}(d,q-1)$. Hence we may suppose that $d | q - 1$.

Now from Lemma 3B,

$$\sum_x \psi(ax^d) = \sum_{\chi \text{ of exp. } d} \bar{\chi}(a)\ G(\chi,\psi) .$$

There are precisely d characters of exponent d. One of them is χ_0 and has $G(\chi_0, \psi) = 0$. The other $d-1$ characters have $G(\chi, \psi)$ of modulus $q^{1/2}$. The theorem follows.

§4. The low road.

As promised in §2, we shall give an easy proof of Theorems 2B, 2B′, using Theorem 2A of Ch. I.

LEMMA 4A. *Suppose g is a generator of (the cyclic group) F_q^*, and $\chi \neq \chi_0$ is a multiplicative character of exponent d. Then*

$$\sum_{k=0}^{d-1} \chi(g^k) = 0 .$$

Proof. χ is a character (but not the principal character) of the factor group $F_q^*/(F_q^*)^d$. On the other hand, $g^0, g^1, \ldots, g^{d-1}$ run through the cosets of this factor group. The lemma thus follows from Theorem 1D.

Proof of Theorem 2B. Again let g be a generator of F_q^*. Let Z_k be the number of x with $f(x)$ in the coset $g^k(F_q^*)^d$. Then

$$\sum_{x \in F_q} \chi(f(x)) = \sum_{k=0}^{d-1} Z_k \chi(g^k) . \tag{4.1}$$

Now let N_k be the number of $(x,y) \in F_q^2$ with

$$y^d = f(x)\, g^{-k} . \tag{4.2}$$

Since $Y^d - f(X)\, g^{-k}$ is again absolutely irreducible by Lemma 2C of Ch. I, it follows that Theorem 2A of Ch. I is applicable and that

$|N_k - q| < 4md^{3/2}q^{1/2}$. Let N_k' be the number of solutions of (4.2) with $y \neq 0$. Then $|N_k' - N_k| \le m$, so that $|N_k' - q| < 5md^{3/2}q^{1/2}$. If we write $Z_k = (q/d) + R_k$ and observe that $Z_k = N_k'/d$, we obtain

$$|R_k| < 5md^{1/2}q^{1/2} .$$

Now (4.1) in conjunction with Lemma 4A yields

$$\left| \sum_{x \in F_q} \chi(f(x)) \right| = \left| \sum_{k=0}^{d-1} \left(\frac{q}{d} + R_k\right) \chi(g^k) \right| = \left| \sum_{k=0}^{d-1} R_k \chi(g^k) \right|$$

$$\le \sum_{k=0}^{d-1} |R_k| < 5md^{3/2}q^{1/2} .$$

<u>LEMMA 4B</u>. <u>Let</u> $f(X)$ <u>be a polynomial in</u> $F_q[X]$, <u>and let</u> d <u>be a divisor of</u> $q-1$. <u>The following three conditions are equivalent.</u>

(i) $f(X) = ck(X)^d$ <u>with</u> $c \in F_q$, $k(X) \in F_q[X]$.

(ii) $f(X) = h(X)^d$ with $h(X) \in \bar{F}_q[X]$.

(iii) $f(X) = c(X - x_1)^{e_1} \cdots (X - x_s)^{e_s}$ <u>with</u> $x_i \in \bar{F}_q$ and $d | e_i$ $(i = 1, \ldots, s)$.

<u>Proof</u>. If (i) holds, then (ii) is true with $h(X) = c^{1/d}k(X)$. Clearly (ii) implies (iii). If (iii) holds, set $k(X) = (X - x_1)^{e_1/d} \cdots (X - x_s)^{e_s/d}$. Then $f(X) = ck(X)^d$, and we have to show that $k(X) \in F_q[X]$. Write $k(X) = X^u + c_1X^{u-1} + \cdots + c_u$. We know that $k(X)^d \in F_q[X]$. The coefficient of X^{du-1} in $k(X)^d$ is dc_1 . Since $d \neq 0$ in F_q , it follows that $c_1 \in F_q$. Suppose we know that $c_1, \ldots, c_{i-1} \in F_q$. The coefficient of X^{du-i} in $k(X)^d$

is dc_i plus a polynomial in $c_1, \ldots, c_{i-1}$ with coefficients in F_q. Hence c_i also is in F_q.

Proof of Theorem 2B'. Write $f(X) = c(X - x_1)^{e_1} \ldots (X - x_s)^{e_s}$ where $x_1, \ldots, x_s$ are distinct elements of $\bar{F}_q$. By our hypothesis, $e = \text{g.c.d.}(e_1, \ldots, e_s, d)$ is a proper divisor of d. We have

$$f(X) = ck(X)^e ,$$

where $k(X) = (X - x_1)^{e_1/e} \ldots (X - x_s)^{e_s/e}$. By Lemma 4B, applied with e in place of d, we see that $k(X) \in F_q[X]$. Since g.c.d. $(e_1/e, \ldots, e_s/e, d/e) = 1$, it follows from Lemma 2C of Ch. I that

$$Y^{d/e} - k(X)$$

is absolutely irreducible. The character χ^e is of exponent d/e and is not the principal character since $e \neq d$. By Theorem 2B,

$$\left| \sum_x \chi(f(x)) \right| = \left| \chi(c) \sum_x \chi^e(k(x)) \right| < 5(m/e)(d/e)^{3/2} q^{1/2} \leq 5md^{3/2} q^{1/2} .$$

§5. Systems of equations $y_1^{d_1} = f_1(x), \ldots, y_n^{d_n} = f_n(x)$.

Throughout, $f_1(X), \ldots, f_n(X)$ will be polynomials with coefficients in F_q and of degree $\leq m$. Put

(5.1) $$\delta = \text{l.c.m.}(d_1, \ldots, d_n) \quad \text{and} \quad d = d_1 d_2 \ldots d_n .$$

THEOREM 5A. *Let* X *be a variable and let* $\eta_1, \ldots, \eta_n$ *be algebraic quantities with*

(5.2) $$\eta_1^{d_1} = f_1(X), \ldots, \eta_n^{d_n} = f_n(X) .$$

Suppose

$$[\bar{F}_q(X,\eta_1, \ldots ,\eta_n): \bar{F}_q(X)] = d . \tag{5.3}$$

Then if $q > 100\,\delta^3 m^2 n^2$, the number N of solutions $(x,y_1, \ldots ,y_n) \in F_q^{n+1}$ of the equations in the title satisfies

$$|N - q| < 5mnd\,\delta^{5/2} q^{1/2} .$$

Proof. Write $d_i' = \text{g.c.d.}(d_i, q-1)$. By an argument used in Ch. I, §2 , the number of solutions of the equations in the title is the same as the number of solutions of

$$y_1^{d_1'} = f_1(x), \ldots ,y_n^{d_n'} = f_n(x) . \tag{5.4}$$

Moreover, write $d_i = d_i' e_i$ and let $\eta_1', \ldots ,\eta_n'$ satisfy $\eta_i'^{d_i'} = f_i(X)$ $(i = 1, \ldots ,n)$, and let $\eta_1, \ldots ,\eta_n$ have $\eta_i^{e_i} = \eta_i'$ $(i = 1, \ldots ,n)$. Then (5.2) and hence (5.3) holds. We have

$$[\bar{F}_q(X,\eta_1', \ldots, \eta_n'): \bar{F}_q(X)] \le d_1' \ldots d_n' ,$$

$$[\bar{F}_q(X,\eta_1, \ldots ,\eta_n): \bar{F}_q(X,\eta_1', \ldots ,\eta_n')] \le e_1 \ldots e_n .$$

Hence in view of (5.3) ,

$$[\bar{F}_q(X,\eta_1', \ldots ,\eta_n'): \bar{F}_q(X)] = d_1' \ldots d_n' = d' ,$$

say. Therefore the system of equations (5.4) also satisfies the hypothesis of the theorem. We may therefore suppose without loss of generality that

$$d_i \mid (q-1) \qquad (i = 1, \ldots ,n) . \tag{5.5}$$

Let χ be a character of order δ, and let χ_i be the character

$$\chi_i = \chi^{\delta/d_i} \qquad (i = 1, \ldots, n) .$$

Then χ_i is of order d_i. The characters of exponent d_i are

$$\chi_i^0 = \chi_o, \chi_i, \chi_i^2, \ldots, \chi_i^{d_i - 1} .$$

By Lemma 2A, the number of $y \in F_q$ with $y^{d_i} = w$ equals

$$\sum_{\chi \text{ of exp. } d_i} \chi(w) = \sum_{j=0}^{d_i - 1} \chi_i^j(w) .$$

Hence

$$N = \sum_{x \in F_q} \sum_{j_1=0}^{d_1-1} \cdots \sum_{j_n=0}^{d_n-1} \chi_1^{j_1}(f_1(x)) \ldots \chi_n^{j_n}(f_n(x))$$

(5.6)
$$= \sum_{j_1=0}^{d_1-1} \cdots \sum_{j_n=0}^{d_n-1} \left(\sum_{x \in F_q} \chi(f_1(x)^{j_1\delta/d_1} \ldots f_n(x)^{j_n\delta/d_n}) \right) .$$

The main term is for $j_1 = \ldots = j_n = 0$, and it equals q. The other summands are character sums

$$\sum_{x \in F_q} \chi(g(x))$$

with

$$g(X) = f_1(X)^{j_1\delta/d_1} \ldots f_n(X)^{j_n\delta/d_n} = \mathfrak{Y}_1^{j_1\delta} \ldots \mathfrak{Y}_n^{j_n\delta}$$

having $j_1, \ldots, j_n \neq 0, \ldots, 0$. If $g(X)$ were a δth power in $\bar{F}_q[X]$, then $\mathfrak{Y}_1^{j_1} \ldots \mathfrak{Y}_n^{j_n} \in \bar{F}_q[X]$. But in view of (5.3), the elements $\eta_1^{j_1} \cdots \eta_n^{j_n}$ with $0 \le j_i < d_i$ $(i = 1, \ldots, n)$ are a field basis of $\bar{F}_q(X, \eta_1, \ldots, \eta_n)$ over $\bar{F}_q(X)$, and hence $\eta_1^{j_1} \cdots \eta_n^{j_n} \notin \bar{F}_q[X]$ if some j_i is not 0. Thus $g(X)$ is not a δth power, i.e., it is not of the type of Lemma 4B with δ in place of d. By Theorem 2B′, and since $q > 100\,\delta^3 m^2 n^2 = 100\,\delta(\delta mn)^2$, we get

$$\left| \sum_{x \in F_q} \chi(g(x)) \right| < 5(mn\delta)\,\delta^{3/2} q^{1/2} .$$

In view of (5.6), we obtain

$$|N - q| < 5mn\,\delta^{5/2}\,dq^{1/2} .$$

Recall that the "big O" notation $O(g(n))$ always stands for a function $f(n)$ with $|f(n)| \le c\,g(n)$ for some fixed $c > 0$.

COROLLARY 5B. *Let* t *be a fixed positive integer. For a prime* p, *let* $L = L_t(p)$ *be the number of* x (mod p) *such that*

$$x+1, x+2, \ldots x+t$$

are (non-zero) quadratic residues mod p. *Then for large* p,

$$L = \frac{p}{2^t} + O(p^{1/2}) .$$

Deduction of the Corollary. In the field F_p, consider the system of equations

$$y_1^2 = x+1, \ldots, y_t^2 = x+t . \tag{5.7}$$

In the notation of Theorem 5A, $m = 1$ and, since $d_i = 2$ for $1 \le i \le t$, we have $d = 2^t$. Let $\mathfrak{Y}_1, \ldots, \mathfrak{Y}_t$ be quantities with

$$\mathfrak{Y}_1^2 = X+1, \ldots, \mathfrak{Y}_t^2 = X+t .$$

In order to apply Theorem 5A to this case, we need that

$$[\bar{F}_p(X, \mathfrak{Y}_1, \ldots, \mathfrak{Y}_t) : \bar{F}_p(X)] = 2^t .$$

This is true if $p \ge t$, $p \ne 2$, as may be shown as an exercise. In fact, the reader might want to do the following

<u>Exercise.</u> Let D be a unique factorization domain of characteristic $\ne 2$ with quotient field K. Let $p_1, \ldots, p_t$ be distinct primes in D. Then

$$[K(\sqrt{p_1}, \ldots, \sqrt{p_t}) : K] = 2^t .$$

(See also Besicovitch (1940)).

If N is the number of solutions of the system (5.7), then by Theorem 5A,

$$|N - p| = O(p^{1/2}) .$$

If N' is the number of solutions with $x+1, \ldots, x+t$ all non-zero, then $|N - N'| = O(1)$, so that

$$|N' - p| = O(p^{1/2}) .$$

Since $N' = 2^t L$, the Corollary follows.

§6. Auxiliary lemmas on $\omega_1^\nu + \dots + \omega_\ell^\nu$.

Given a complex valued function $f(\nu)$ and a real valued function $g(\nu) > 0$, the Vinogradov notation

$$f(\nu) \ll g(\nu)$$

means that $|f(\nu)| < c\, g(\nu)$ for some positive constant c and for $\nu = 1, 2, \dots$. Thus it means that $f(\nu) = O(g(\nu))$.

LEMMA 6A. *Let* $\omega_1, \dots, \omega_\ell$ *be complex numbers, and let* $B > 0$. *If*

$$\omega_1^\nu + \dots + \omega_\ell^\nu \ll B^\nu \quad \textit{for} \quad \nu = 1, 2, \dots, \tag{6.1}$$

then $|\omega_j| \le B$ $(j = 1, \dots, \ell)$.

Proof. For small values of $|z|$, we have

$$-\log(1 - \omega z) = \omega z + \frac{1}{2}\omega^2 z^2 + \frac{1}{3}\omega^3 z^3 + \dots .$$

Thus

$$-\log((1 - \omega_1 z) \dots (1 - \omega_\ell z)) = \sum_{\nu=1}^{\infty} \frac{1}{\nu}(\omega_1^\nu + \dots + \omega_\ell^\nu) z^\nu . \tag{6.2}$$

In view of (6.1), the sum on the right is convergent for $|z| < B^{-1}$. Hence the function (6.2) is analytic for $|z| < B^{-1}$. Thus $1 - \omega_j z \ne 0$ if $|z| < B^{-1}$, and therefore $|\omega_j| \le B$ $(j = 1, \dots, \ell)$.

In our proof we used facts about analytic functions. We now shall prove a stronger result without using analytic functions. Write $\Re z$ for the real part of z.

LEMMA 6B. *Let* $\omega_1, \dots, \omega_\ell$ *be complex numbers, and let* $B > 0$,

$c > 0$. *If*

(6.3) $$\Re(\omega_1^\nu + \ldots + \omega_\ell^\nu) < B^\nu \quad (\nu = 1,2, \ldots),$$

then $|\omega_j| \leq B \quad (j = 1, \ldots, \ell)$.

This is an immediate consequence of the even stronger

LEMMA 6C. *Let* $\omega_1, \ldots, \omega_\ell$ *be non-zero complex numbers. There are infinitely many positive integers* ν *with*

(6.4) $$\Re(\omega_1^\nu + \ldots + \omega_\ell^\nu) > (1 - 2\pi\nu^{-1/\ell})(|\omega_1|^\nu + \ldots + |\omega_\ell|^\nu),$$

hence with

$$\Re(\omega_1^\nu + \ldots + \omega_\ell^\nu) > (1 - \varepsilon)(|\omega_1|^\nu + \ldots + |\omega_\ell|^\nu),$$

for given $\varepsilon > 0$.

For the proof we shall need Dirichlet's Theorem on Simultaneous Approximations:

LEMMA 6D. *Let* $\theta_1, \ldots, \theta_\ell$ *be real. There exist* $(\ell+1)$*-tuples of integers* $\nu, m_1, \ldots, m_\ell$ *with arbitrarily large* $\nu > 0$ *and with*

(6.5) $$\left|\theta_i - \frac{m_i}{\nu}\right| < \nu^{-1-(1/\ell)} \quad (i = 1, \ldots, \ell).$$

Proof. Write $\alpha = [\alpha] + \{\alpha\}$, where $[\alpha]$ is the *integer part of* α, i.e. the integer with $\alpha - 1 < [\alpha] \leq \alpha$, and where $\{\alpha\}$ is the *fractional part of* α, i.e., the number with $0 \leq \{\alpha\} < 1$ such that $\alpha - \{\alpha\}$ is an integer.

Now suppose $N > 0$ is an integer. The points

$$(6.6) \qquad (\{u\theta_1\}, \ldots, \{u\theta_\ell\})$$

with $u = 0,1, \ldots, N^\ell$ are $N^\ell + 1$ points in the half open unit cube $0 \le x_1 < 1, \ldots, 0 \le x_\ell < 1$. This unit cube may be decomposed in an obvious way onto N^ℓ half open small cubes of side N^{-1}. Two of the points (6.6) will lie in the same small cube. If these points belong to the parameters u', u with $u' < u$, then

$$|\{u\theta_j\} - \{u'\theta_j\}| < N^{-1} \qquad (j = 1, \ldots, \ell),$$

or

$$|u\theta_j - u'\theta_j - m_j| < N^{-1} \qquad (j = 1, \ldots, \ell)$$

for certain integers $m_1, \ldots, m_\ell$. Putting $\nu = u - u'$, we have

$$(6.7) \qquad |\nu\theta_j - m_j| < N^{-1} \qquad (j = 1, \ldots, \ell),$$

whence (6.5) in view of $\nu \le N^\ell$.

If at least one of the θ_j is irrational, then as $N \to \infty$, the inequalities (6.7) cannot be satisfied with bounded values of ν. Hence there will be $(\ell+1)$-tuples with (6.5) and with arbitrarily large values of ν. If all the θ_j are rational, say if $\theta_j = a_j/b$ $(j = 1, \ldots, \ell)$ with $b > 0$, we may set

$$\nu = tb, \quad m_1 = ta_1, \ldots, m_\ell = ta_\ell$$

with $t = 1,2, \ldots$.

Proof of Lemma 6C. Observe that for real θ , η

$$|e(\theta) - e(\eta)| \le 2\pi|\theta - \eta| .$$

Write $\omega_j = |\omega_j| e(\theta_j)$ with real θ_j . There will be infinitely many ν , and integers $m_1, \dots, m_\ell$, having

$$|\nu\theta_j - m_j| < \nu^{-1/\ell} \qquad (j = 1, \dots, \ell) .$$

For such ν ,

$$|e(\nu\theta_j) - 1| = |e(\nu\theta_j) - e(m_j)| \le 2\pi|\nu\theta_j - m_j| < 2\pi\nu^{-1/\ell}$$

whence

$$\Re(\omega_j^\nu) = |\omega_j|^\nu \Re(e(\nu\theta_j)) > (1 - 2\pi\nu^{-1/\ell})|\omega_j|^\nu \quad (j = 1, \dots, \ell) ,$$

whence (6.4).

§7. Further auxiliary lemmas.

LEMMA 7A.. Let ν , m be positive integers. Writing (ν,m) = g.c.d.(ν,m) , we have the polynomial identity

$$(7.1) \qquad \prod_{u=1}^{\nu} (1 - e(mu/\nu)X) = (1 - X^{\frac{\nu}{(\nu,m)}})^{(\nu,m)} .$$

Proof. In the case $(\nu,m) = 1$, the identity reduces to

$$\prod_{u=1}^{\nu} (1 - e(mu/\nu)X) = 1 - X^\nu .$$

It is correct in this case, since both sides are polynomials of degree ν with constant term 1 and with roots $e(-mu/\nu)$ $(u = 1, \dots , \nu)$

In general, put $\nu = \nu_1(\nu,m)$, $m = m_1(\nu,m)$. As u runs through a residue system modulo ν , it runs (ν,m) times through a residue system modulo ν_1 . Thus (7.1) is obtained by raising

$$\prod_{u=1}^{\nu_1} (1 - e(m_1 u/\nu_1)X) = 1 - X^{\nu_1} \tag{7.2}$$

to the $(\nu,m)^{th}$ power. But (7.2) is correct by the special case already considered, since $(\nu_1, m_1) = 1$.

LEMMA 7B. *Let* $h(X) = X^d + a_1 X^{d-1} + \ldots + a_d$ *be an irreducible polynomial in* $F_q[X]$. *Then in* $F_{q^\nu}[X]$ *it splits into* $r = (\nu, d)$ *irreducible polynomials of degree* d/r :

$$h(X) = h_1(X) \ldots h_r(X) . \tag{7.3}$$

If we normalize $h_i(X)$ *such that its leading coefficient is* 1, *then* $h_i(X) \in F_{q^r}[X]$ $(i = 1, \ldots, r)$. *The elements* σ *of the Galois group of* F_{q^r}/F_q *permute*[†)] *the polynomials* $h_1, \ldots, h_r$. *Given* h_i, h_j, *there is a* σ *in the Galois group with* $\sigma h_i = h_j$.

Proof. Consider the fields

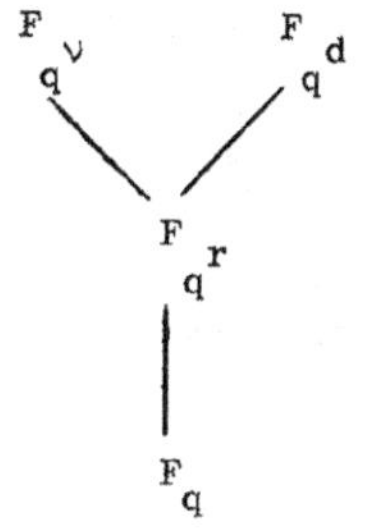

The roots of $h(X)$ are algebraic of degree d over F_q, hence lie in F_{q^d}. They are algebraic of degree d/r over F_{q^r}. Hence in F_{q^r}, the polynomial $h(X)$ has the factorization (7.3), where each h_i is of degree d/r and is irreducible over F_{q^r}. Since

---

†) we let σ operate on the *coefficients* of the polynomials.

$(d/r, \nu/r) = 1$, the roots are still of degree d/r over F_{q^ν}, and hence the polynomials $h_i(X)$ are still irreducible over F_{q^ν}. The elements σ of the Galois group $G = G(F_{q^r}/F_q)$ leave h invariant; hence they permute $h_1, \ldots, h_r$. Given i in $1 \leq i \leq r$, the polynomial

$$\prod_{\sigma \in G} \sigma h_i$$

is invariant under G, hence lies in $F_q[X]$. It has roots in common with the irreducible polynomial h, hence equals $h = h_1 \ldots h_r$. So as σ runs through G, then σh_i runs through $h_1, \ldots, h_r$.

§8. Zeta Function and L-Functions.

Throughout, $h = h(X)$ will denote a monic (i.e. with leading coefficient 1) polynomial with coefficients in F_q. If $h(X)$ is of degree d, put

$$\underline{\underline{\mathfrak{N}}}(h) = q^d .$$

For complex

$$s = \sigma + it ,$$

put

$$\zeta(s) = \sum_h \frac{1}{\underline{\underline{\mathfrak{N}}}(h)^s} .$$

Here the sum is over monic polynomials $h \in F_q[X]$.

THEOREM 8A. (i) The sum for $\zeta(s)$ is absolutely convergent for $\sigma > 1$, in fact uniformly convergent for $\sigma > \sigma_0 > 1$.

(ii) For $\sigma > 1$,

$$\zeta(s) = \prod_{h \text{ irred.}} (1 - \underline{\underline{\mathfrak{N}}}(h)^{-s})^{-1},$$

where the product is over irreducible monic polynomials in $F_q[X]$.

(iii) $$\zeta(s) = \frac{1}{1-q^{1-s}}.$$

Proof. (i)

$$(8.1) \qquad \zeta(s) = \sum_{d=0}^{\infty} \frac{N(d)}{q^{ds}} = \sum_{d=0}^{\infty} \frac{q^d}{q^{ds}},$$

where $N(d)$ is the number of monic polynomials of degree d. The sum on the right is clearly absolutely convergent if $\sigma > 1$, and uniformly so if $\sigma > \sigma_0 > 1$.

(ii) Since every polynomial may uniquely be written as a product of powers of irreducible polynomials, we have, for $\sigma > 1$,

$$\zeta(s) = \prod_{h \text{ irred.}} \left(1 + \frac{1}{\underline{\underline{\mathfrak{N}}}(h)^s} + \frac{1}{\underline{\underline{\mathfrak{N}}}(h^2)^s} + \dots\right)$$

$$= \prod_{h \text{ irred.}} (1 - \underline{\underline{\mathfrak{N}}}(h)^{-s})^{-1}.$$

(iii) follows immediately from (8.1).

Remark. We call $\zeta(s)$ a Zeta Function. It is almost (but not quite) the Zeta Function of the "function field" $F_q(X)$. For a reader familiar with Zeta Functions of function fields, we remark the following. The prime divisors of the rational function field $F_q(X)$ consist of prime divisors which correspond to irreducible monic polynomials, plus the "infinite" prime divisor. Our Zeta Function differs from the Zeta Function of the field $F_q(X)$ in that in the product (ii) the factor corresponding to the infinite prime divisor is missing. This is why we have $\zeta(s) = (1-q^{1-s})^{-1}$, while the Zeta Function of the function field is $(1-q^{-s})^{-1}(1-q^{1-s})^{-1}$.

Let G be the group of rational functions $h_1(X)/h_2(X)$, where h_1, h_2 are monic in $F_q[X]$. Let $\bar{G}$ be a subgroup of G such that

(8.2) $$\text{if } h_1h_2 \in \bar{G}\text{, then } h_1, h_2 \in \bar{G}$$

for polynomials h_1, h_2. Let χ be a character on $\bar{G}$. We extend the definition of χ by setting $\chi(h) = 0$ if h is a polynomial not in $\bar{G}$. Then still $\chi(h_1h_2) = \chi(h_1)\chi(h_2)$ for monic polynomials h_1, h_2. For $s = \sigma + it$, put

$$L(s,\chi) = \sum_h \chi(h)\,\underline{\underline{\mathfrak{N}}}(h)^{-s},$$

where the sum is over monic polynomials $h \in F_q[X]$.

THEOREM 8B. (i) The sum for $L(s,\chi)$ is absolutely convergent for $\sigma > 1$, in fact uniformly convergent for $\sigma > \sigma_0 > 1$.

(ii) For $\sigma > 1$,

$$L(s,\chi) = \prod_{h \text{ irred.}} (1-\chi(h)\underline{\underline{\mathfrak{N}}}(h)^{-s})^{-1}.$$

Proof. Everything works almost the same as in parts (i), (ii) of Theorem 8A. The details are left as an exercise.

Remark. The experts will see that our functions $L(s,\chi)$ are L-Functions associated with the function field $F_q(X)$.

§9. Special L-Functions.

Let $f(X)$ be a fixed monic polynomial in $F_q[X]$. In $\bar{F}_q[X]$ it factors into $(X+\gamma_1)^{a_1} \dots (X+\gamma_m)^{a_m}$, say. Let $\bar{G}$ be the subgroup of G consisting of rational functions $r(X) = h_1(X)/h_2(X)$ having $h_1(\gamma_i)h_2(\gamma_i) \neq 0$ $(i = 1, \dots, m)$. Then $\bar{G}$ satisfies (8.2). For $r(X) \in \bar{G}$, put[†)]

(9.1) $$\{r\} = r(\gamma_1)^{a_1} \dots r(\gamma_m)^{a_m} .$$

If $r(X) = (X+\alpha_1) \dots (X+\alpha_u)(X+\beta_1)^{-1} \dots (X+\beta_v)^{-1}$, then

$$\{r\} = f(\alpha_1) \dots f(\alpha_u) f(\beta_1)^{-1} \dots f(\beta_v)^{-1} .$$

Always $\{r\} \in F_q$ and $\{r_1 r_2\} = \{r_1\}\{r_2\}$. Thus if χ is a multiplicative character of F_q, then

$$\chi(\{r_1 r_2\}) = \chi(\{r_1\})\chi(\{r_2\}) .$$

Therefore $\chi(\{r\})$ is a character on the group $\bar{G}$.

Let $\bar{\bar{H}}$ be the subgroup of $\bar{G}$ consisting of $r(X) = h_1(X)/h_2(X)$ with $h_1(\gamma_i) = h_2(\gamma_i) \neq 0$ $(i = 1, \dots, m)$.

LEMMA 9A. $\chi(\{r\}) = 1$ for $r \in \bar{\bar{H}}$.

Proof. Obvious.

Let $g(X)$ be a fixed polynomial in $F_q[X]$, of degree n and with constant term zero. Given $r = r(X) \in G$, put $[r] = 0$ if $r(X) = 1$, and

---

†) Put $\{r\} = 1$ if $f(X) = 1$.

*) We allow $n = 0$, $g(X) = 0$.

(9.2) $$[r] = g(\alpha_1)+\dots+g(\alpha_u) - g(\beta_1)-\dots-g(\beta_v)$$

if $r(X) = (X+\alpha_1)\dots(X+\alpha_u)(X+\beta_1)^{-1}\dots(X+\beta_v)^{-1}$ with $\alpha_1,\dots,\alpha_u$, $\beta_1,\dots,\beta_v$ in $\bar{F}_q$. Then $[r] \in F_q$ and $[r_1r_2] = [r_1]+[r_2]$. Thus if ψ is an additive character of F_q, then

$$\psi([r_1r_2]) = \psi([r_1])\psi([r_2]) .$$

Thus $\psi([r])$ is a character on the group G.

Let H be the subset of G consisting of rational functions $r(X) = h_1(X)/h_2(X)$ having

(9.3) $$h_1(X) = X^u + a_1X^{u-1}+\dots+a_u , \quad h_2(X) = X^v + b_1X^{v-1}+\dots+b_v$$

with

(9.4) $$a_1 = b_1 , \quad a_2 = b_2, \dots, a_n = b_n$$

For example, polynomials X^u lie in H, and so do polynomials $X^u + a_{n+1}X^{u-n-1}+\dots+a_u$ with $u > n$. It is easily seen that H is a subgroup of G.

<u>LEMMA 9B</u>. $\psi([r]) = 1$ <u>if</u> $r \in H$.

<u>Proof</u>. In (9.2), $g(\alpha_1)+\dots+g(\alpha_u)$ is a symmetric polynomial of degree n in $\alpha_1,\dots,\alpha_u$. Hence it is a polynomial in the first n elementary symmetric polynomials in $\alpha_1,\dots,\alpha_u$, i.e., in the coefficients $a_1,\dots,a_n$ in (9.3):

$$g(\alpha_1)+\dots+g(\alpha_u) = \ell_1(a_1,\dots,a_n)$$

with a polynomial ℓ_1. Similarly,

$$g(\beta_1)+\dots+g(\beta_v) = \ell_2(b_1,\dots,b_n) .$$

Since the two symmetric functions $g(\alpha_1)+\dots+g(\alpha_u)$ and $g(\beta_1)+\dots+g(\beta_v)$ have constant term zero, and since they are "the same", except perhaps for the number of variables, the two polynomials ℓ_1 and ℓ_2 are the same. Thus (9.4) implies that $[r] = 0$, whence $\psi([r]) = 1$.

Now put

$$\mathcal{X}(r) = \chi(\{r\})\psi([r]) .$$

Then $\mathcal{X}$ will be a character on the group $\bar{G}$. Let $\bar{H}$ be the intersection $\bar{H} = H \cap \bar{\bar{H}}$. Then $\bar{H}$ is a subgroup of $\bar{G}$, and we have the

COROLLARY 9C. $\mathcal{X}(r) = 1$ if $r \in \bar{H}$.

LEMMA 9D. Suppose $\ell \geq 0$. Then every coset of $\bar{H}$ in $\bar{G}$ contains precisely q^ℓ polynomials of degree $n + m + \ell$.

Proof. It will suffice to show that if $r(X)$ is in $\bar{G}$, then there are precisely q^ℓ polynomials $k(X) = X^{n+m+\ell} + b_1X^{n+m+\ell-1} + \dots + b_{n+m+\ell}$ with $k(X)/r(X) \in \bar{H}$. If $r(X)$ has the expansion $r(X) = X^u + a_1X^{u-1} + a_2X^{u-2} + \dots$, then this condition means that

(9.5) $$b_1 = a_1,\dots,b_n = a_n$$

and that

(9.6) $$k(\gamma_i) = r(\gamma_i) \qquad (i = 1,\dots,m) .$$

The coefficients $b_1,\dots,b_n$ are determined by (9.5). Pick

$b_{n+1}, \ldots, b_{n+\ell}$ arbitrary. Then the relations (9.6) are m (non-homogeneous) linear equations in the m remaining coefficients $b_{n+\ell+1}, \ldots, b_{n+\ell+m}$. The matrix of this system of equations is (γ_i^j) $(1 \le i \le m\ ,\ 0 \le j \le m-1)$. The determinant is a Van der Monde determinant. Since $\gamma_1, \ldots, \gamma_m$ are distinct, the determinant is non-zero. Thus we can solve the system (9.6) uniquely.

Hence our freedom consists precisely in picking $b_{n+1}, \ldots, b_{n+\ell}$. This gives q^ℓ possibilities.

LEMMA 9E. *Suppose that*

(9.7) *either* $\chi \ne \chi_o$ *is of exponent* d *and* $Y^d - f(X)$ *is absolutely irreducible,*
or $\psi \ne \psi_o$ *and either* (i) $(n,q) = 1$ *or, more generally,* (ii) $Z^q - Z - g(X)$ *is absolutely irreducible.*

Then the character X *is not principal, i.e.,* $X(k) \ne 1$ *for some* $k \in \bar{G}$.

Proof. Suppose $X(k) = 1$. Then $\chi(\{k\})\psi([k]) = 1$. Since $\chi(\{k\})$ is a dth root of unity and $\psi([k])$ is a pth root of unity with $(d,p) = 1$, it is easily seen that

$$\chi(\{k\}) = \psi([k]) = 1 .$$

Hence in the first case of the lemma it will suffice to find a k with $\chi(\{k\}) \ne 1$, and in the second case it will suffice to find a k with $\psi([k]) \ne 1$.

If $\chi \ne \chi_o$, suppose it to be of order e with $e \mid d$. Since $Y^d - f(X)$ is absolutely irreducible, not all the exponents in

$f(X) = (X+\gamma_1)^{a_1} \ldots (X+\gamma_m)^{a_m}$ are multiples of e. (See Lemma 2C of Ch. I). Say $e \nmid a_1$. Given $c_2, \ldots, c_m$ in F_q^*, we can therefore pick $c_1 \in F_q^*$ with $c_1^{a_1} \ldots c_m^{a_m} \notin (F_q^*)^e$, whence with $\chi(c_1^{a_1} \ldots c_m^{a_m}) \neq 1$. By the argument of Lemma 9D, there is a polynomial $k(X) \in \bar{G}$ with

$$k(\gamma_i) = c_i \qquad (i = 1, \ldots, m) .$$

Then $\{k\} = c_1^{a_1} \ldots c_m^{a_m}$ and $\chi(\{k\}) \neq 1$.

If $\psi \neq \psi_o$, suppose first, (i), that $(n,q) = 1$ and $f(X) = 1$. Say, $g(X) = aX^n + g_1(X)$ where g_1 is of degree $< n$. If $k(X) = X^n + v = (X+\alpha_1) \ldots (X+\alpha_n)$, then $g_1(\alpha_1) + \ldots + g_1(\alpha_n) = 0$, since it is a polynomial with constant term zero in the first $n-1$ elementary symmetric polynomials in $\alpha_1, \ldots, \alpha_n$. On the other hand, $\alpha_1^n + \ldots + \alpha_n^n = (-1)^{n+1} nv$, so that $[k] = (-1)^{n+1} anv$ and

$$\psi([k]) = \psi([X^n + v]) = \psi((-1)^{n+1} anv) . \tag{9.8}$$

For a proper choice of v, $\psi([k]) \neq 1$, since n is not divisible by the characteristic.

More generally, (ii), let $\psi \neq \psi_o$, and let $Z^q - Z - g(X)$ be absolutely irreducible. For every $b \in F_q^*$, $bZ^q - bZ - g(X) = (bZ)^q - (bZ) - g(X)$ is absolutely irreducible. So for $a \in F_q^*$, also $Z^q - Z - ag(X)$ is absolutely irreducible, and hence

$$Z^p - Z - ag(X) \tag{9.9}$$

is[†)] absolutely irreducible, where p is the characteristic.

Write $\mathfrak{T}$, $\mathfrak{T}_\nu$, $\mathfrak{T}'_\nu$, respectively, for the trace $F_q \to F_p$, $F_{q^\nu} \to F_q$, $F_{q^\nu} \to F_p$. The character ψ is of the type $\psi(z) =$

---

†) For if $q = p^\nu$, then $Z^q - Z = u(Z)^p - u(Z)$ with $u(Z) = Z^{p^{\nu-1}} + \ldots + Z^p + Z$.

$\psi_a(z) = e(\mathfrak{T}(az)/p)$ for some $a \in F_q^*$. If N_ν is the number of zeros (x,z) of (9.9) in F_{q^ν}, then by Theorem 1A of Ch. III, $N_\nu = q^\nu + O(q^{\nu/2})$. Hence if ν is large, $N_\nu < pq^\nu$. Now for given $x \in F_{q^\nu}$, either $\mathfrak{T}'_\nu(ag(x)) = 0$, in which case by Lemma 1F of Ch. I there are p values of $z \in F_{q^\nu}$ with $z - z - ag(x) = 0$. Or $\mathfrak{T}'_\nu(ag(x)) \neq 0$, in which case there is no such z. Since $N_\nu < pq^\nu$, there will be an $x \in F_{q^\nu}$ with $\mathfrak{T}'_\nu(ag(x)) \neq 0$. Put $k(X) = (X + x_1) \dots (X + x_\nu) \in F_q[X]$, where $x_1 = x, \dots, x_\nu$ are the conjugates of x over F_q. Then $[k] = g(x_1) + \dots + g(x_\nu) = \mathfrak{T}_\nu(g(x))$ and

$$\psi([k]) = e(\mathfrak{T}(a\mathfrak{T}_\nu g(x)))/p) = e(\mathfrak{T}\mathfrak{T}_\nu(ag(x))/p)$$

$$= e(\mathfrak{T}'_\nu(ag(x))/p) \neq 1 .$$

By the freedom in the choice of x we may ensure that $k \in \bar{G}$.

LEMMA 9F. Suppose the hypothesis (9.7) of Lemma 9E holds. Suppose $\ell \geq 0$. Then

$$\sum_{\substack{h \in G \\ h \text{ monic pol.} \\ \deg h = n+m+\ell}} \chi(h) = 0 .$$

Proof. By Corollary 9C and by Lemma 9E, χ induces a non-principal character on the finite factor group $\bar{G}/\bar{H}$. On the other hand, as h runs through polynomials of $\bar{G}$ of degree $n+m+\ell$, then by Lemma 9D, it will lie precisely q^ℓ times in every given coset of $\bar{G}/\bar{H}$. The lemma is therefore a consequence of Theorem 1D.

As in §8, extend the definition of χ by putting $\chi(h) = 0$ if h is a polynomial $\notin \bar{G}$. As in §8, form the L-Function $L(s,\chi)$.

<u>THEOREM 9G. Again suppose (9.7). Putting</u> $U = q^{-s}$, <u>we have</u>

$$L(s,\chi) = 1 + c_1 U + \dots + c_{n+m-1} U^{n+m-1} \tag{9.9}$$

<u>If</u> $\chi \neq \chi_0$ <u>or if</u> $\chi = \chi_0$, $f(X) = 1$, <u>then</u>

$$c_1 = \sum_{x \in F_q} \chi(f(x))\psi(g(x)) .$$

<u>Proof.</u> $L(s,\chi) = 1 + c_1 U + c_2 U^2 + \dots$ with

$$c_t = \sum_{\substack{h \in \bar{G} \\ \text{pol. of deg. } t}} \chi(h) .$$

Here $c_t = 0$ if $t \geq n + m$, by Lemma 9F. Hence $L(s,\chi)$ is a polynomial in U of degree $< n + m$. Now

$$c_1 = \sum_{\substack{h \in \bar{G} \\ \deg h=1}} \chi(h) = \sum_{\substack{x \\ x+\gamma_i \neq 0}} \chi(X + x)$$

$$= \sum_{\substack{x \\ x+\gamma_i \neq 0}} \chi(\{X + x\})\psi([X + x])$$

$$= \sum_{\substack{x \\ x+\gamma_i \neq 0}} \chi((\gamma_1 + x)^{a_1} \dots (\gamma_m + x)^{a_m})\ \psi(g(x))$$

$$= \sum_{\substack{x \\ f(x) \neq 0}} \chi(f(x))\psi(g(x))$$

$$= \sum_{x} \chi(f(x))\psi(g(x)) .$$

§10. Field extensions. The Hasse-Davenport relations.

Given an overfield F_{q^ν} of F_q, write $\mathfrak{N}_\nu$ for the norm from F_{q^ν} to F_q and $\mathfrak{T}_\nu$ for the trace from F_{q^ν} to F_q. If χ is a multiplicative character of F_q, then χ_ν defined by

$$\chi_\nu(x) = \chi(\mathfrak{N}_\nu(x))$$

is a multiplicative character of F_{q^ν}. If ψ is an additive character of F_q, then ψ_ν defined by

$$\psi_\nu(x) = \psi(\mathfrak{T}_\nu(x))$$

is an additive character of F_{q^ν}.

As in §9, let $f(X) \in F_q[X]$ be monic, with a factorization $(X + \gamma_1)^{a_1} \ldots (X + \gamma_m)^{a_m}$ in $\bar{F}_q[X]$. Let G_ν be the group of rational functions $r(X) = h_1(X)/h_2(X)$ with monic $h_i(X) \in F_{q^\nu}[X]$ $(i = 1,2)$, and let $\bar{G}_\nu$ be the subgroup consisting of rational functions having $h_1(\gamma_i)h_2(\gamma_i) \neq 0$ $(i = 1, \ldots, m)$. For $r(X) \in \bar{G}_\nu$, define $\{r\}$ by (9.1). Then $\chi_\nu(\{r\})$ will be a character on $\bar{G}_\nu$. The definition of $\bar{\bar{H}}_\nu$ is now obvious, and the obvious analog of Lemma 9A holds.

Again, let $g(X) \in F_q[X]$ be of degree n and with constant term zero. For $r = r(X) \in G_\nu$, define $[r]$ by (9.2). Then $\psi_\nu([r])$ will be a character on G_ν; the analog of Lemma 9B holds, if H_ν is defined in the obvious way.

It is now clear that

$$\mathrm{X}_\nu(r) = \chi_\nu(\{r\})\psi_\nu([r]) \tag{10.1}$$

is a character on $\bar{G}_\nu$, which is 1 for $r(X) \in \bar{H}_\nu = H_\nu \cap \bar{\bar{H}}_\nu$.

The sum we are interested in is

(10.2) $$S = \sum_{x \in F_q} \chi(f(x))\psi(g(x)) \ ;$$

we now put

(10.3) $$S_\nu = \sum_{x \in F_{q^\nu}} \chi_\nu(f(x))\psi_\nu(g(x)) \ .$$

We put

$$L_\nu(s,\mathcal{X}) = \sum_{h \in F_{q^\nu}[X]} \mathcal{X}_\nu(h)\underline{\underline{\mathfrak{N}}}_\nu(h)^{-s} \ ,$$

where $\underline{\underline{\mathfrak{N}}}_\nu(h) = q^{\nu d}$ if $d = \deg h$. The main result of this section is

THEOREM 10A.

$$L_\nu(s,\mathcal{X}) = \prod_{u=1}^{\nu} L\left(s - \frac{2\pi i u}{\nu \log q}, \mathcal{X}\right) \ .$$

Before proving this theorem, we note the following supplement to Lemma 7B:

LEMMA 10B. *Make the same assumptions as in Lemma 7B, and let (7.3) be the factorization of* $h(X)$ *in* $F_{q^\nu}[X]$. *Then*

(i) $\underline{\underline{\mathfrak{N}}}_\nu(h_i) = \underline{\underline{\mathfrak{N}}}(h)^{\nu/r}$,

(ii) $\mathcal{X}_\nu(h_i) = \mathcal{X}(h)^{\nu/r}$.

Proof. (i) $\underline{\underline{\mathfrak{N}}}_\nu(h_i) = q^{\nu(d/r)} = \underline{\underline{\mathfrak{N}}}(h)^{\nu/r}$.

(ii) We have $\{h\} = \{h_1\} \dots \{h_r\}$. Here by Lemma 7B , $\{h_1\}, \dots, \{h_r\}$ are in F_{q^r} and are conjugates over F_q . Hence if $\mathfrak{N}_r$ is the norm from F_{q^r} to F_q , then $\{h\} = \mathfrak{N}_r(\{h_i\})$ $(i = 1, \dots, r)$. Thus[†)]

$$\mathfrak{N}_\nu(\{h_i\}) = \mathfrak{N}_r(\{h_i\})^{\nu/r} = \{h\}^{\nu/r} \quad (i = 1, \dots, r) . \tag{10.4}$$

On the other hand, $[h] = [h_1] + \dots + [h_r]$. Therefore $[h] = \mathfrak{T}_r([h_i])$ $(i = 1, \dots, r)$, where $\mathfrak{T}_r$ is the trace from F_{q^r} to F_q . Thus

$$\mathfrak{T}_\nu([h_i]) = \frac{\nu}{r}\mathfrak{T}_r([h_i]) = \frac{\nu}{r}[h] \quad (i = 1, \dots, r) . \tag{10.5}$$

In view of the definition of χ_ν as given in (10.1), the desired conclusion follows from (10.4), (10.5).

Proof of Theorem 10A. By the product formula of Theorem 8B,

$$L_\nu(s, \chi) = \sum_{\substack{\ell \text{ irred, monic} \\ \text{in } F_{q^\nu}[X]}} (1 - \chi_\nu(\ell)\underline{\underline{\mathfrak{N}}}_\nu(\ell)^{-s})^{-1} .$$

An irreducible monic polynomial $h(X) \in F_q[X]$ of degree d splits over F_{q^ν} according to Lemmas 7B, 10C into $h(X) = \ell_1(X) \dots \ell_r(X)$ with $r = \text{g.c.d.}(d, \nu)$ and with $\chi_\nu(\ell_i)\underline{\underline{\mathfrak{N}}}_\nu(\ell_i)^{-s} = (\chi(h)\underline{\underline{\mathfrak{N}}}(h)^{-s})^{\nu/r}$.

On the other hand, every monic irreducible $\ell(X) \in F_{q^\nu}[X]$ is the

---

†) Observe that $\underline{\underline{\mathfrak{N}}}_\nu$ is defined on polynomials with coefficients in F_{q^ν} , and is quite distinct from $\mathfrak{N}_\nu$, the norm from F_{q^ν} to F_q .

factor of a unique monic irreducible $h(X) \in F_q[X]$. Therefore

$$L_\nu(s,\chi) = \prod_{\substack{h \text{ irred, monic,} \\ \text{in } F_q[X]}} \left(1 - (\chi(h)\underline{\underline{\mathfrak{N}}}(h)^{-s})^{\nu/(\nu, \deg h)}\right)^{-(\nu,\, \deg h)}$$

Applying Lemma 7A with $m = \deg h$ and $X = \chi(h)\underline{\underline{\mathfrak{N}}}(h)^{-s}$, we obtain

$$L_\nu(s,\chi) = \prod_{\substack{h \text{ irred, mon.} \\ \text{in } F_q[X]}} \prod_{u=1}^{\nu} (1 - e(u \deg h/\nu)\chi(h)\underline{\underline{\mathfrak{N}}}(h)^{-s})^{-1}$$

$$= \prod_{u=1}^{\nu} \prod_{\substack{h \text{ irred, mon.} \\ \text{in } F_q[X]}} (1 - \chi(h)\underline{\underline{\mathfrak{N}}}(h)^{-(s-(2\pi iu/(\nu \log q)))})$$

$$= \prod_{u=1}^{\nu} L(s - \frac{2\pi iu}{\nu \log q}, \chi) .$$

Recall that under the condition (9.7), $L(s,\chi)$ was a polynomial in $U = q^{-s}$ with constant term 1 (see Theorem 9G). Thus it is of the form $(1 - \omega_1 U) \ldots (1 - \omega_k U)$ with complex $\omega_1, \ldots, \omega_k$. We now have the

COROLLARY 10C. *If* $L(s,\chi)$ *is given by*

$$L(s,\chi) = (1 - \omega_1 U) \ldots (1 - \omega_k U)$$

with $U = q^{-s}$, *then*

$$L_\nu(s,\chi) = (1 - \omega_1^\nu U_\nu) \ldots (1 - \omega_k^\nu U_\nu)$$

with $U_\nu = q^{-\nu s}$

Proof.

$$q^{-(s-(2\pi iu/(\nu \log q)))} = e(u/\nu)U ,$$

so that

$$L(s-(2\pi iu/(\nu \log q)),\chi) = (1-\omega_1 e(u/\nu)U) \ldots (1-\omega_k e(u/\nu)U) .$$

Thus by Theorem 10A,

$$L_\nu(s,\chi) = \left(\prod_{u=1}^{\nu}(1-\omega_1 e(u/\nu)U)\right) \ldots \left(\prod_{u=1}^{\nu}(1-\omega_k e(u/\nu)U)\right)$$

$$= (1-\omega_1^\nu U^\nu) \ldots (1-\omega_k^\nu U^\nu)$$

$$= (1-\omega_1^\nu U_\nu) \ldots (1-\omega_k^\nu U_\nu).$$

COROLLARY 10D. Suppose that (9.7) holds. Suppose that $\chi \neq \chi_0$ or $\chi = \chi_0$ with $f(X) = 1$. Then the sum S_ν given by (10.3) is of the form

$$S_\nu = -\omega_1^\nu - \ldots - \omega_{n+m-1}^\nu .$$

Proof. By Theorem 9G, applied to F_{q^ν} instead of F_q, and by Corollary 10C,

$$L_\nu(s,\chi) = 1 + c_{\nu,1}U_\nu + \ldots + c_{\nu,n+m-1}U_\nu^{n+m-1}$$

$$= (1-\omega_1^\nu U_\nu) \ldots (1-\omega_{n+m-1}^\nu U_\nu) ,$$

with

$$c_{\nu,1} = S_\nu \ .$$

On the other hand, it is clear that $c_{\nu,1} = -(\omega_1^\nu + \ldots + \omega_{n+m-1}^\nu)$, and the corollary follows.

<u>COROLLARY 10E</u>. (Davenport-Hasse Relation). <u>Let</u> χ, ψ <u>be a multiplicative and an additive character of</u> F_q. <u>Recall that the Gaussian sum</u> $G(\chi,\psi)$ <u>was</u> $\sum_x \chi(x)\psi(x)$, <u>over</u> $x \in F_q$. <u>Now put</u>

$$G_\nu(\chi,\psi) = \sum_{x \in F_{q^\nu}} \chi_\nu(x)\psi_\nu(x) \ .$$

<u>Then unless</u> $\chi = \chi_0$, $\psi = \psi_0$ <u>and</u> ν <u>is even</u>,

$$-G_\nu(\chi,\psi) = (-G(\chi,\psi))^\nu \ .$$

See Davenport - Hasse (1935).

<u>Proof</u>. Suppose $\chi \neq \chi_0$. We have $G(\chi,\psi) = S$ and $G_\nu(\chi,\psi) = S_\nu$ where S, S_ν are given by (10.2), (10.3) with $f(X) = g(X) = X$. Thus $n = m = 1$. By Corollary 10D, $S_\nu = -\omega_1^\nu$ for $\nu = 1,2,\ldots$, whence $S_\nu = -(-S_1)^\nu$. The case when $\chi = \chi_0$ follows from (3.1), (3.3).

§11. Proof of the Principal Theorems.

(a) Theorems 2C, 2C′. We deal with multiplicative character sums. So let $\chi \neq \chi_0$ be a multiplicative character, and let $\psi = \psi_0$. Let $f(X)$ be as in Theorem 2C and monic, and put $g(X) = 0$, so that $n = \deg g = 0$. In this case

(11.1) $$S = \sum_{x \in F_q} \chi(f(x)) \quad \text{and} \quad S_\nu = \sum_{x \in F_{q^\nu}} \chi_\nu(f(x)) \ .$$

In view of Corollary 10D ,

(11.2) $$S_\nu = -\omega_1^\nu - \dots - \omega_{m-1}^\nu .$$

Now suppose that χ is of exponent d where $d > 1$ and $d \mid q-1$. There are d characters χ of exponent d . For each such character χ , we may define the sums $S = S_\chi$ and $S_\nu = S_{\chi\nu}$. We then have for $\chi \neq \chi_0$,

(11.3) $$S_{\chi\nu} = -\omega_{\chi 1}^\nu - \dots - \omega_{\chi, m-1}^\nu .$$

Taking the sum over $\chi \neq \chi_0$ of exponent d , we obtain

(11.4) $$\sum_{\substack{\chi \neq \chi_0 \\ \text{of exp. } d}} S_{\chi\nu} = - \sum_{\substack{\chi \neq \chi_0 \\ \text{of exp. } d}} \sum_{i=1}^{m-1} \omega_{\chi i}^\nu .$$

On the other hand, for $\chi = \chi_0$, (11.1) yields

(11.5) $$S_{\chi_0\nu} = q^\nu .$$

<u>LEMMA 11A</u>. <u>For given</u> $w \in F_{q^\nu}$, <u>the number of</u> $y \in F_{q^\nu}$ <u>with</u> $y^d = w$ <u>equals</u>

$$\sum_{\substack{\chi \\ \text{of exp. } d}} \chi_\nu(w) = \sum_{\substack{\chi \\ \text{of exp. } d}} \chi(\mathfrak{N}_\nu(w)) .$$

<u>Proof</u>. We first note that the map $w \to \mathfrak{N}(w)$ is a group homomorphism $F_{q^\nu}^* \to F_q^*$. For each $z \in F_q^*$, the number of $w \in F_{q^\nu}^*$ with

$$\mathfrak{N}(w) = w^{1+q+\dots+q^{\nu-1}} = z$$

is $\le 1+q+\dots+q^{\nu-1} = (q^\nu-1)/(q-1) = |F^*_{q^\nu}|/|F^*_q|$; hence the number of these w is exactly this number, and our homomorphism is onto. The restriction of the map to $(F^*_{q^\nu})^d$ is a map $(F^*_{q^\nu})^d \to (F^*_q)^d$, and comparing cardinalities we see that it is onto again.

According to Lemma 2A, the sum in Lemma 11A is d or 0 or 1 , respectively, if $\mathfrak{N}(w) \in (F^*_q)^d$ or $\notin (F^*_q)^d$, $\ne 0$, or $= 0$. In the first case, by what we just said, $w \in (F^*_{q^\nu})^d$, and there are d elements y with $y^d = w$. In the second case, $w \notin (F^*_{q^\nu})^d$, $\ne 0$ and there are no solutions y with $y^d = w$. In the third case, $w = 0$, and there is the single solution $y = 0$.

Writing N_ν for the number of solutions x,y in F_{q^ν} of $y^d = f(x)$, we immediately obtain

LEMMA 11B.

$$N_\nu = \sum_{\chi \text{ of exp. } d} \sum_x \chi_\nu(f(x)) = \sum_{\chi \text{ of exp. } d} S_{\chi\nu} .$$

Now we know from Theorem 2A of Ch. I that if $Y^d - f(X)$ is absolutely irreducible, then

$$N_\nu - q^\nu \ll q^{\nu/2} . \tag{11.6}$$

Combining this with (11.4), (11.5) and Lemma 11B, we obtain

$$\sum_{\substack{\chi \ne \chi_0 \\ \chi \text{ of exp. } d}} \sum_{i=1}^{m-1} \omega^\nu_{\chi i} \ll q^{\nu/2} .$$

Lemma 6A yields

(11.7) $$|\omega_{\chi i}| \leq q^{1/2}$$

for all χ , i under consideration. Thus from (11.2) or (11.3), $|S_\nu| \leq (m-1)q^{\nu/2}$, and $|S| \leq (m-1)q^{1/2}$.

We assumed that $f(X)$ was monic. But since $\chi(af(x)) = \chi(a)\chi(f(x))$, our character sum estimate clearly holds in general. Therefore the proof of Theorem 2C is complete. Theorem 2C′ can be deduced from Theorem 2C in the same way in which Theorem 2B′ was deduced from Theorem 2B.

We remark that Lemma 11B, together with (11.4), (11.5), (11.7) and the fact that there are $d-1$ characters $\chi \neq \chi_0$ of exponent d , gives

$$|N_\nu - q^\nu| \leq (d-1)(m-1)q^{\nu/2} ,$$

and

$$|N-q| \leq (d-1)(m-1)q^{1/2} .$$

This improves upon Theorem 2A of Ch. I.

(b) Theorem 2E. We next consider additive character sums. So let $\psi \neq \psi_0$ be an additive character, and let $\chi = \chi_0$. Let $g(X)$ be as in Theorem 2E, and put $f(X) = 1$, so that in the notation of §9, 10, $m = \deg f = 0$. In this case

(11.8) $$S = \sum_{x \in F_q} \psi(g(x)) \quad \text{and} \quad S_\nu = \sum_{x \in F_{q^\nu}} \psi_\nu(g(x)) .$$

By Corollary 10D,

$$S_\nu = -\omega_1^\nu - \dots - \omega_{n-1}^\nu .$$

There are q additive characters ψ of F_q . For each such ψ we may define $S_\nu = S_{\psi\nu}$, and for each $\psi \neq \psi_0$, we have

(11.9) $$S_{\psi\nu} = -\omega_{\psi 1}^{\nu} - \cdots - \omega_{\psi, n-1}^{\nu} .$$

Taking the sum over characters $\psi \neq \psi_0$, we get

(11.10) $$\sum_{\psi \neq \psi_0} S_{\psi\nu} = -\sum_{\psi \neq \psi_0} \sum_{i=1}^{n-1} \omega_{\psi i}^{\nu} .$$

On the other hand, for $\psi = \psi_0$, (11.8) yields

(11.11) $$S_{\psi_0 \nu} = q^{\nu} .$$

<u>LEMMA 11C.</u> <u>For given</u> $w \in F_{q^\nu}$, <u>the number of</u> $z \in F_{q^\nu}$ <u>with</u> $z^q - z = w$ <u>equals</u>

(11.12) $$\sum_{\psi} \psi_\nu(w) = \sum_{\psi} \psi(\mathfrak{T}_\nu(w)) .$$

<u>Proof</u>. We shall use Theorem 1F of Ch. I. If $\mathfrak{T}(w) = 0$, then on the one hand, we have q solutions $z \in F_{q^\nu}$ of $z^q - z = w$, and on the other hand, our sum (11.12) is q by Theorem 1D. If $\mathfrak{T}(w) \neq 0$, then there is no $z \in F_{q^\nu}$ with $z^q - z = w$, and the sum (11.12) is zero, by Theorem 1D again.

Writing N_ν for the number of solutions x, z in F_{q^ν} of $z^q - z = g(x)$, we obtain

<u>LEMMA 11D.</u>

$$N_\nu = \sum_{\psi} \sum_{x} \psi_\nu(g(x)) = \sum_{\psi} S_{\psi\nu}$$

Now suppose we know somehow that (11.6) holds. Then very much as in (a), we may conclude that

$$(11.13) \qquad |\omega_{\psi i}| \le q^{1/2} \qquad (\psi \ne \psi_0 \ ; \ i = 1, \ldots, n-1) \ ,$$

and hence that $|S| \le (n-1)q^{1/2}$.

Now if condition (i) of Theorem 2E holds, then (11.6) is true by Theorem 9A of Ch. I. Or if (ii), $Z^q - Z - g(X)$ is absolutely irreducible, then (11.6) is true by Theorem 1A of Ch. III.

We assumed that $g(X)$ had constant term zero. Now since $\psi(g(x) + a) = \psi(a)\psi(g(x))$, it is clear that the modulus of the character sum S does not change if we replace $g(X)$ by $g(X) + a$. On the other hand, the hypotheses of Theorem 2E are not affected by this change. This is obvious for (i). As for (ii), we note that every a is of the type $a = b^q - b$ for some $b \in \bar{F}_q$, and hence $Z^q - Z - g(X) - a = (Z - b)^q - (Z - b) - g(X)$ is absolutely irreducible if and only if $Z^q - Z - g(X)$ is. Thus Theorem 2E is completely proved.

We remark that in view of (11.10), (11.11), (11.13) and Lemma 11D, we have

$$(11.14) \qquad |N_\nu - q^\nu| \le (q-1)(n-1)q^{\nu/2} \ ,$$

which is an improvement upon Theorem 9A of Ch. I.

(c) Theorem 2G. Suppose $f(X)$, $g(X)$ satisfy the hypotheses of Theorem 2G. Assume initially that $f(X)$ is monic and that $g(X)$ has constant term zero. For every multiplicative character χ of exponent d and every additive character ψ , we put

$$S_{\chi\psi\nu} = \sum_{x \in F_{q^\nu}} \chi_\nu(f(x))\psi_\nu(g(x)) \ .$$

By Corollary 10D,

(11.15) $$S_{\chi\psi\nu} = -(\omega^{\nu}_{\chi\psi 1} + \dots + \omega^{\nu}_{\chi\psi, m+n-1}) \quad (\chi \neq \chi_o \text{ of exp. } d,\ \psi \neq \chi_o)$$

On the other hand, by (11.3),

$$S_{\chi\nu} = S_{\chi\psi_o\nu} = -(\omega^{\nu}_{\chi 1} + \dots + \omega^{\nu}_{\chi, m-1}) \qquad (\chi \neq \chi_o) .$$

Also, by (11.9),

$$S_{\psi\nu} = S_{\chi_o\psi\nu} = -(\omega^{\nu}_{\psi 1} + \dots + \omega^{\nu}_{\psi, n-1}) \qquad (\psi \neq \psi_o) .$$

Finally,

$$S_{\chi_o\psi_o\nu} = q^{\nu} .$$

<u>LEMMA 11E.</u> <u>For</u> $w_1, w_2 \in F_{q^\nu}$ <u>the number of</u> $y, z \in F_{q^\nu}$ <u>with</u>

$$y^d = w_1, \qquad z^q - z = w_2$$

<u>is</u>

$$\sum_{\substack{\chi \\ \text{of exp } d}} \sum_{\psi} \chi_\nu(w_1)\psi_\nu(w_2)$$

<u>Proof.</u> Combine Lemmas 11A, 11C .

We obtain

<u>LEMMA 11F.</u> <u>The number</u> N_ν <u>of</u> x,y,z <u>in</u> F_{q^ν} <u>with</u> $y^d = f(x)$, $z^q - z = g(x)$ <u>is given by</u>

$$N_\nu = \sum_{\substack{\chi \\ \text{of exp } d}} \sum_{\psi} S_{\chi\psi\nu} .$$

Now suppose we know from some source that (11.6) holds. Then

$$|\omega_{\chi\psi i}| \le q^{1/2} \quad \text{for } \chi \text{ of exp. } d,\ (\chi,\psi) \ne (\chi_0,\psi_0), \text{ and} \tag{11.16}$$

$$i = 1, \ldots, m+n-1 .$$

In view of (11.15), we obtain $|S_\nu| \le (m+n-1)q^{\nu/2}$ and $|S| \le (m+n-1)q^{1/2}$.

Now under the conditions of Theorem 2G, the equations $y^d = f(x)$, $z^q - z = g(x)$ define an absolute curve. (See Example 3 in §2 of Ch. VI). So (11.6) holds by Theorem 7A of Ch. VI.

Theorem 2G is proved, since the restrictions that $f(X)$ be monic and $g(X)$ be of constant term zero, can be easily removed.

§12. Kloosterman Sums.

It is easily seen that the sum (2.3) is -1 if $a \ne 0$, $b = 0$, or if $a = 0$, $b \ne 0$; hence we may suppose that $ab \ne 0$.

LEMMA 12A. Let q be odd and let $\chi(x)$ be the quadratic character of F_q , i.e., $\chi(x) = 1$ or $\chi(x) = -1$ if $x \ne 0$ is a square or a non-square in F_q , respectively; and $\chi(0) = 0$. Then if $\psi \ne \psi_0$ and if $ab \ne 0$,

$$\sum_{x \in F_q^*} \psi(ax + bx^{-1}) = \sum_{x \in F_q} \psi(x)\chi(x^2 - 4ab) . \tag{12.1}$$

Proof. The sum on the left hand side is

$$\sum_{y \in F_q} \psi(y)Z(y) , \tag{12.2}$$

where $Z(y)$ is the number of $x \in F_q^*$ with $y = ax + bx^{-1}$. Solving this equation for x we obtain $x = (2a)^{-1}(y \pm \sqrt{y^2 - 4ab})$, which may or may not lie in F_q. We obtain

$$Z(y) = \chi(y^2 - 4ab) + 1 :$$

For if $y^2 - 4ab \neq 0$ is a square (or a non-square), then $Z(y) = 2$ (or 0); and if $y^2 - 4ab = 0$, then $Z(y) = 1$. Thus (12.2) becomes

$$\sum_y \psi(y)\chi(y^2 - 4ab) + \sum_y \psi(y) = \sum_x \psi(x)\chi(x^2 - 4ab) .$$

The polynomials $Y^2 - (X^2 - 4ab)$ and $Z^q - Z - X$ are absolutely irreducible. Hence by Theorem 2G, the sum on the right hand side of (12.1) has modulus $\le (m + n - 1)q^{1/2} = (2 + 1 - 1)q^{1/2} = 2q^{1/2}$.

This completes the proof of Theorem 2H if q is odd, but it depends on Theorem 2G, which in turn depends on Ch. VI. But we needed Ch. VI only to show (11.6), i.e., $N_\nu - q^\nu \ll q^{\nu/2}$. But in our case the number N_ν is the number of solutions x,y,z in F_{q^ν} of

$$y^2 = x^2 - 4ab , \quad z^q - z = x .$$

This number N_ν is also the number of solutions y,z of

$$y^2 = (z^q - z)^2 - 4ab .$$

Since $Y^2 - (Z^q - Z)^2 + 4ab$ is absolutely irreducible, the number N_ν satisfies (11.6) by Theorem 2A of Ch. I.

We now will sketch another proof of Theorem 2H, which works for q even as well. Let G again be the group of rational functions $h_1(X)/h_2(X)$ whose numerators and denominators are monic polynomials. Let $\hat{G}$ be the subgroup of functions whose numerators and denominators have non-zero constant term. Given $r(X) \in \hat{G}$, put $[r] = 1$ if $r(X) = 1$, and

$$[r] = a(\alpha_1 + \dots + \alpha_u - \beta_1 - \dots - \beta_v) + b\left(\frac{1}{\alpha_1} + \dots + \frac{1}{\alpha_u} - \frac{1}{\beta_1} - \dots - \frac{1}{\beta_v}\right)$$

if $r(X) = (X+\alpha_1)\dots(X+\alpha_u)(X+\beta_1)^{-1}\dots(X+\beta_v)^{-1}$ with $\alpha_1,\dots,\alpha_u$, $\beta_1,\dots,\beta_v$ in $\overline{F_q}$. Then $[r] \in F_q$ and $[r_1 r_2] = [r_1] + [r_2]$.

The function

$$\chi(r) = \psi([r])$$

will be a character on $\hat{G}$. Let $\hat{H}$ be the subset of $\hat{G}$ consisting of $r(X) = h_1(X)/h_2(X)$ having

$$h_1(X) = X^u + a_1 X^{u-1} + \dots + a_{u-1}X + a_u, \quad h_2(X) = X^v + b_1 X^{v-1} + \dots + b_{v-1}X + b_v$$

with

$$a_1 = b_1, \quad \frac{a_{u-1}}{a_u} = \frac{b_{v-1}}{b_v}.$$

For example, monomials X^u lie in $\hat{H}$, and so do polynomials of degree $u \geq 2$ of the type $X^u + a_2X^{u-2} + \dots + a_{u-2}X^2 + a_u$. It is easily seen that $\hat{H}$ is a subgroup of $\hat{G}$. As an analog of Lemma 9B, we now observe

LEMMA 12B. $\chi(r) = 1$ if $r \in \hat{H}$.

Proof. If $r \in \hat{H}$, then

$$\alpha_1 + \dots + \alpha_u - \beta_1 - \dots - \beta_v = a_1 - b_1 = 0 ,$$

$$\frac{1}{\alpha_1} + \dots + \frac{1}{\alpha_u} - \frac{1}{\beta_1} - \dots - \frac{1}{\beta_v} = \frac{a_{u-1}}{a_u} - \frac{b_{v-1}}{b_v} = 0 ,$$

so that $[r] = 0$.

The analog of Lemma 9D is

LEMMA 12C. Suppose $\ell \geq 0$. Then every coset of $\hat{H}$ in $\hat{G}$ contains precisely $q^{\ell}(q-1)$ polynomials of degree $\ell + 3$.

The proof of this is left as an exercise. Carrying out the obvious analog to the argument in §9, one sees that the L-Function $L(s,\chi)$ is a polynomial in $U = q^{-s}$ of the type

$$L(s,U) = 1 + c_1 U + c_2 U^2 = (1 - \omega_1 U)(1 - \omega_2 U)$$

with

$$c_1 = \sum_{x \in F_q^*} \psi(ax + bx^{-1}) .$$

Thus it suffices to show that $|\omega_i| \leq q^{1/2}$ $(i = 1,2)$. This is accomplished by showing that the number N_ν of solutions x, z in F_{q^ν} of $x \neq 0$, $z^q - z = ax + bx^{-1}$, satisfies (11.6). Since clearly $aX^2 - (Z^q - Z)X + b$ is absolutely irreducible, this follows from Theorem 1A of Ch. III.

§13. Further Results.

Let $\psi = \psi_o$ be an additive character. Let $g(X)$ be a polynomial of degree n with $(n,q) = 1$ and with constant term zero. We know from Theorem 9G that if $\chi(r) = \psi([r])$, where $[r]$ is defined as in §9, then with $U = q^{-s}$,

$$L(s,\chi) = 1 + c_1 U + \ldots + c_{n-1} U^{n-1} .$$

We now prove

THEOREM 13A. $|c_{n-1}| = q^{(n-1)/2}$.

Proof. We have

$$c_{n-1} = \sum_{\substack{h \text{ monic} \\ \deg h = n-1}} \chi(h) ,$$

so that

$$|c_{n-1}|^2 = \sum_{\substack{h_1 \\ \deg n-1}} \sum_{\substack{h_2 \\ \deg n-1}} \chi(h_1/h_2) .$$

Now $\chi(k)$ depends only on the coset C of k modulo the subgroup H of G. Thus

(13.1) $$|c_{n-1}|^2 = \sum_C \chi(C) Z(C) ,$$

where the sum is over cosets C of H in G, and where $Z(C)$ is the number of pairs of monic polynomials h_1, h_2 of degree $n-1$ with $h_1/h_2 \in C$.

We write $r_1 \equiv r_2 \pmod{H}$ if $r_1/r_2 \in H$, i.e., if r_1, r_2 lie in the same coset C. If we expand the rational functions as

$$r_i(X) = X^{u_i} + a_{i1}X^{u_i-1} + a_{i2}X^{u_i-2} + \dots \quad (i = 1,2) ,$$

then $r_1 \equiv r_2 \pmod{H}$ if and only if

$$a_{11} = a_{21}, \dots, a_{1n} = a_{2n}$$

Thus if $C(v_1, \dots, v_n)$ consists of rational functions $r(X) = X^u + a_1X^{u-1} + \dots$ with $a_1 = v_1, \dots, a_n = v_n$, then the sets $C(v_1, \dots, v_n)$ are just the cosets of H in G.

Now h_1/h_2 with $h_1 = X^{n-1} + a_1X^{n-2} + \dots + a_{n-1}$, $h_2 = X^{n-1} + b_1X^{n-2} + \dots + b_{n-1}$ lies in $C(v_1, \dots, v_n)$ precisely if

$$
\begin{aligned}
a_1 &= b_1 + v_1 \\
a_2 &= b_2 + b_1v_1 + v_2 , \\
&\vdots \\
a_{n-1} &= b_{n-1} + b_{n-2}v_1 + \dots + b_1v_{n-2} + v_{n-1} , \\
0 &= b_{n-1}v_1 + \dots + b_1v_{n-1} + v_n .
\end{aligned}
\tag{13.2}
$$

Thus $Z(C(v_1, \dots, v_n))$ is simply the number of solutions in $a_1, \dots, a_{n-1}$, $b_1, \dots, b_{n-1}$ in F_q of (13.2).

<u>LEMMA 13B.</u>

$$Z(C(v_1, \dots, v_n)) = \begin{cases} q^{n-2} & \underline{\text{if}} \ v_1, \dots, v_{n-1} \ \underline{\text{are not}} \ 0, \dots, 0 , \\ q^{n-1} & \underline{\text{if}} \ v_1 = \dots = v_{n-1} = v_n = 0 , \\ 0 & \underline{\text{if}} \ v_1 = \dots = v_{n-1} = 0 , \ v_n \neq 0 . \end{cases}$$

Proof. The number of solutions $a_1, \ldots, a_{n-1}$, $b_1, \ldots, b_{n-1}$ in F_q of (13.2) is just the number of solutions $b_1, \ldots, b_{n-1}$ in F_q of the last equation (13.2).

In view of (9.8), we have

LEMMA 13C. $\chi(C(0, \ldots, 0, v)) = \psi((-1)^{n+1} n \alpha v)$, where a is the leading coefficient of $g(X)$.

The proof of Theorem 13A is now completed as follows. By (13.1), Lemmas 13B, 13C, and since $\chi(C(0, \ldots, 0)) = 1 = \psi(0)$, we obtain

$$|c_{n-1}|^2 = \sum_{v_1} \cdots \sum_{v_n} \chi(C(v_1, \ldots, v_n)) Z(C(v_1, \ldots, v_n))$$

$$= q^{n-2} \sum_{v_1} \cdots \sum_{v_n} \chi(C(v_1, \ldots, v_n))$$

$$+ (q^{n-1} - q^{n-2}) \chi(C(0, \ldots, 0))$$

$$- q^{n-2} \sum_{v \neq 0} \chi(C(0, \ldots, 0, v)) .$$

Here the first summand is zero, since $C(v_1, \ldots, v_n)$ ranges through all the cosets of H in G. Combining the second and third summand, we obtain

$$q^{n-1} - q^{n-2} \sum_{v} = \psi((-1)^{n+1} n \alpha v) = q^{n-1} .$$

The proof of Theorem 13A is complete. Now we know that in $L(s, \chi) = (1 - \omega_1 U) \ldots (1 - \omega_{n-1} U)$, the absolute values $|\omega_j| \leq q^{1/2}$ $(j = 1, \ldots, n-1)$. But in view of Theorem 13A, we now have

(13.3) $$|\omega_j| = q^{1/2} \quad (j = 1, \ldots, n-1) .$$

COROLLARY 13D. *Let* $g(X)$ *be of degree* n *with* $(n,q) = 1$, *and let* $\psi \neq \psi_0$ *be an additive character of* F_q . *Then*

$$S_\nu = \sum_{x \in F_{q^\nu}} \psi_\nu(g(x))$$

is of the form

$$S_\nu = -\omega_1^\nu - \ldots - \omega_{n-1}^\nu$$

where $\omega_1, \ldots, \omega_{n-1}$ *have* (13.3).

In particular, neither the exponent $\frac{1}{2}$ nor the constant factor $n-1$ in Theorem 2E may be improved. In fact, by Lemma 6C, we have

COROLLARY 13E. *Let* S_ν *be as above. There are infinitely many positive integers with*

$$|S_\nu| > (n-1)q^{\nu/2}(1 - 2\pi\nu^{-1/(n-1)}) .$$

Similarly, neither the exponent $\frac{1}{2}$ nor the factor $(q-1)(n-1)$ in (11.14) may be improved.

The arguments of this section may be carried over, with suitable changes, to multiplicative character sums and hybrid sums.

III. Absolutely Irreducible Equations $f(x,y) = 0$.

References: Stepanov (1972b, 1974), Schmidt (1973).

§1. Introduction. This chapter is devoted to a proof of

THEOREM 1A. *Suppose* $f(X,Y) \in F_q[X,Y]$ *is absolutely irreducible and of total degree* $d > 0$. *Let* N *be the number of zeros of* f *in* F_q^2 . *If* $q > 250d^5$, *then*

$$(1.1) \qquad |N - q| < \sqrt{2}\, d^{5/2} q^{1/2} .$$

As is well known, this estimate follows from the Riemann Hypothesis for curves over finite fields, which was first proved by Weil (1940, 1948a). In fact, the Riemann Hypothesis gives the stronger estimate

$$|N - q| \leq (d - 1)(d - 2)q^{1/2} + c(d)$$

for some constant $c(d)$. Special cases of Theorem 1A (but with $\sqrt{2}\, d^{5/2}$ replaced by some other constant depending on d) were proved by Stepanov by elementary methods; his most general result was in (1972b, 1974). Stepanov's method was extended by Schmidt (1973) to yield Theorem 1A, and also by Bombieri (1973).

In order to provide easy examples of absolutely irreducible polynomials $f(X,Y)$, we now state

THEOREM 1B. *Let*

$$f(X,Y) = g_0 Y^d + g_1(X)Y^{d-1} + \ldots + g_d(X) ,$$

where g_0 is a non-zero constant, be a polynomial with coefficients in a field k . *Put*

$$\psi(f) = \max_{1 \le i \le d} \frac{1}{i} \deg g_i \quad ,$$

and suppose that $\psi(f) = m/d$ *with* $(m,d) = 1$. *Then* $f(X,Y)$ *is irreducible, in fact absolutely irreducible.*

Remark. The polynomials considered by Stepanov (1972b, 1974) were all of the type of this theorem.

To prove Theorem 1B, we need

LEMMA 1C: If

(1.2) $$f(X,Y) = u(X,Y)\, v(X,Y) \quad ,$$

then $\psi(f) = \max\{\psi(u), \psi(v)\}$.

Proof: Suppose $a + b = d$ and

$$u(X,Y) = u_0 Y^a + u_1(X) Y^{a-1} + \dots + u_a(X) \quad ,$$

$$v(X,Y) = v_0 Y^b + v_1(X) Y^{b-1} + \dots + v_b(X) \quad .$$

Then

$$g_i(X) = \sum_{j+k=i} u_j(X)\, v_k(X) \qquad (0 \le i \le d) \ .$$

Since each summand $u_j(X)\, v_k(X)$ has degree at most $j\psi(u) + k\psi(v) \leqq (j+k)\max(\psi(u),\psi(v)) = i \max(\psi(u),\psi(v))$, we have

$$\frac{1}{i} \deg g_i(X) \le \max\{\psi(u),\psi(v)\} \quad (1 \le i \le d) \ ,$$

whence

$$\psi(f) \leq \max\{\psi(u), \psi(v)\} \quad . \tag{1.3}$$

Now make the substitution $Y \to Y^{\psi}$, where $\psi = \psi(f)$. Then (1.2) becomes

$$g_0 Y^{\psi d} + g_1(X)\, Y^{\psi(d-1)} + \ldots + g_d(X) = (u_0 Y^{\psi a} + u_1(X)\, Y^{\psi(a-1)} + \ldots + u_a(X))$$

$$(v_0 Y^{\psi b} + v_1(X)\, Y^{\psi(b-1)} + \ldots + v_b(X))$$

$$= \hat{u}(X,Y)\, \hat{v}(X,Y) \quad ,$$

say. Examining the total degrees of both sides of this equation[†)], we notice that the L.H.S. has degree ψd , while

$$\deg \hat{u}(X,Y) \geq \psi a \quad \text{and} \quad \deg \hat{v}(X,Y) \geqq \psi b \quad ,$$

so that the R.H.S. has degree $\geqq \psi a + \psi b = \psi d$. Hence in fact

$$\deg \hat{u}(X,Y) = \psi a \quad \text{and} \quad \deg \hat{v}(X,Y) = \psi b \quad .$$

It follows that

$$\deg u_j(X) \leq j\psi \quad (1 \leqq j \leqq a) \quad \text{and} \quad \deg v_k(X) \leq k\psi \quad (1 \leqq k \leqq b) \quad ,$$

whence

$$\psi(u) \leq \psi \quad \text{and} \quad \psi(v) \leq \psi \quad .$$

This, in conjunction with (1.3), proves the lemma.

---

†) It clearly does not matter that our exponents and degrees are not necessarily integers.

Proof of Theorem 1B. Suppose

$$f(X,Y) = u(X,Y)\, v(X,Y)$$

is a proper factorization of $f(X,Y)$. Then

$$\deg_Y u(X,Y) < d \quad \text{and} \quad \deg_Y v(X,Y) < d .$$

We have

$$\psi(u) = \max_{1 \le i \le \deg_Y u} \frac{1}{i} \deg u_i(X) = \frac{r}{s}, \text{ with } 1 \le s < d ,$$

and $$\psi(v) = \max_{1 \le j \le \deg_Y v} \frac{1}{j} \deg v_j(X) = \frac{w}{z}, \text{ with } 1 \le z < d .$$

Hence $\psi(f) \ne \max\{\psi(u), \psi(v)\}$, and the contradiction is obtained by applying Lemma 1C.

The remainder of this section will be used to obtain a very modest reduction of Theorem 1A to a special case.

Suppose $f(X,Y) = g(X,Y^p)$ where, as usual, $q = p^\kappa$. Since $y \to y^p$ is an automorphism of F_q, as (x,y) ranges over all pairs in F_q^2, so does (x,y^p). Therefore the number of zeros of $g(X,Y)$ is equal to the number of zeros of $f(X,Y)$, and we may replace f by g. This process decreases the degree in Y of the polynomial under consideration. After a finite number of such steps, we obtain a polynomial which is not a polynomial in Y^p, i.e. a polynomial which is "separable in Y".

If

$$f(X,Y) = \sum a_{ij} X^i Y^j$$

is separable in Y, then there is some coefficient

$$a_{i_0 j_0} \neq 0 \,, \quad \text{where} \quad p \nmid j_0 \;.$$

Set

$$h(X,Y) = f(X + cY, Y) = \sum a_{ij}(c) X^i Y^j \,.$$

Then the coefficients $a_{ij}(c)$ are polynomials in c of degree at most d, with the properties that

(i) the polynomial $a_{i_0 j_0}(c)$ is not identically zero,

(ii) the coefficient of Y^d is $a_{0d}(c) = f_d(c,1)$,

where $f_d(X,Y)$ consists of the terms of $f(X,Y)$ which are of total degree d. In particular, $a_{0d}(c)$ is not identically zero. If $q > 2d$, (which is the case in Theorem 1A), we can choose $c \in F_q$ so that

$$a_{i_0 j_0}(c) \neq 0 \quad \text{and} \quad a_{0d}(c) \neq 0 \;.$$

Then in the polynomial $h(X,Y)$, Y^d occurs with a non-zero coefficient; moreover, h is separable in Y. Dividing by an appropriate constant, we achieve the following

Reduction: Without loss of generality, we may assume that

$$f(X,Y) = Y^d + g_1(X) Y^{d-1} + \ldots + g_d(X), \quad \deg g_i(X) \leq i \,,$$

and that $f(X,Y)$ is separable in Y.

§2. Independence results.

We begin with a simple remark. Suppose $f(X,Y)$ is a polynomial with coefficients in a field K and of degree $d > 0$ in Y. Suppose $f(X,Y)$ is irreducible over K. Then if we regard $f(X,Y)$ as a polynomial in Y with coefficients in the field $K(X)$, it is still irreducible[†]. Hence if $\mathfrak{Y}$ satisfies $f(X,\mathfrak{Y}) = 0$, then $[K(X,\mathfrak{Y}) : K(X)] = d$.

LEMMA 2A. *Suppose $f(X,Y)$ and $g(Z,U)$ are polynomials with coefficients in a field K, both absolutely irreducible over K. Suppose f is of degree $d > 0$ in Y and g is of degree $d' > 0$ in U. Let $\mathfrak{Y}, \mathfrak{U}$ be quantities with*

$$f(X,\mathfrak{Y}) = 0, \quad g(Z,\mathfrak{U}) = 0 .$$

(So that $[K(X,\mathfrak{Y}) : K(X)] = d$ and $[K(Z,\mathfrak{U}) : K(Z)] = d'$.) Then

$$[K(X,Z,\mathfrak{Y},\mathfrak{U}) : K(X,Z)] = dd' .$$

Remark: The absolute irreducibility of f and g is essential. By way of example, take $K = \mathbb{Q}$ and

---

[†] For suppose to the contrary that $f(X,Y) = g_1(X,Y)g_2(X,Y)$, where the g_i are polynomials of positive degree in Y with coefficients in $K(X)$. Given any polynomial $g(X,Y)$ in Y with coefficients in $K(X)$, we may uniquely write $g(X,Y) = (u(X)/v(X))\,\hat{g}(X,Y)$, where $u(X)$, $v(X)$ are coprime polynomials with leading coefficient 1, and where $\hat{g}(X,Y) = c_0(X) + c_1(X)Y + \dots + c_t(X)Y^t$ with coprime polynomials $c_0(X),\dots,c_t(X)$. Write $r(g) = u(X)/v(X)$. Since $K[X]$ has unique factorization, it can be shown that $r(g_1g_2) = r(g_1)\,r(g_2)$. (This is similar to Gauss' Lemma.) Now if the polynomial $f(X,Y)$ above is irreducible over K, we have $r(f) = 1$, whence $r(g_1)\,r(g_2) = 1$. Thus $f(X,Y) = r(g_1)r(g_2)\hat{g}_1(X,Y)\hat{g}_2(X,Y) = \hat{g}_1(X,Y)\hat{g}_2(X,Y)$ with polynomials $\hat{g}_1$, $\hat{g}_2$, contradicting the irreduciblity of f.

$$f(X,Y) = Y^4 - 2X^2 ,$$

$$g(Z,U) = U^4 - 2Z^2 .$$

If $\mathfrak{Y}$ and $\mathfrak{U}$ are as above, then we have the following diagram:

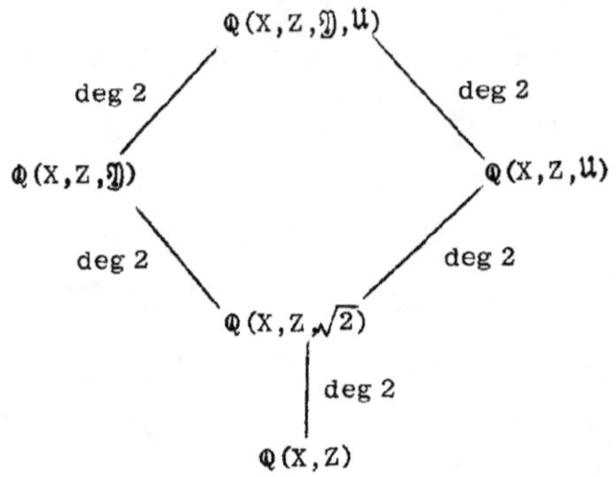

Hence

$$[\mathbb{Q}(X,Z,\mathfrak{Y},\mathfrak{U}) : \mathbb{Q}(X,Z)] = 8 \neq 16 .$$

<u>Proof of the lemma</u>: We need to show that

(2.1) $$[K(X,Z,\mathfrak{Y},\mathfrak{U}) : K(X,Z,\mathfrak{Y})] = d'$$

and

(2.2) $$[K(X,Z,\mathfrak{Y}) : K(X,Z)] = d .$$

To show (2.1) it will suffice to show that $g(Z,U)$ remains irreducible over $K(X,\mathfrak{Y})$. Otherwise,

$$g(Z,U) = g_1(Z,U)\, g_2(Z,U) ,$$

where $g_i(Z,U)$ $(i = 1,2)$ has coefficients in $K(X,\mathfrak{Y})$ and is of degree less than d' in U. Write

$$g_i(Z,U) = \sum_{j,k} c_{ijk} Z^j U^k \qquad (i = 1,2) ,$$

where

$$c_{ijk} = r^{(0)}_{ijk}(X) + r^{(1)}_{ijk}(X)\mathfrak{Y} + \dots + r^{(d-1)}_{ijk}(X)\mathfrak{Y}^{d-1} ,$$

with rational functions $r^{(h)}_{ijk}(X)$.

Pick $x \in \bar{K}$ such that the denominators of the $r^{(h)}_{ijk}(x)$ are non-zero and such that if

$$f(X,Y) = a_0(X)Y^d + a_1(X)Y^{d-1} + \dots + a_d(X) ,$$

then $a_0(x) \neq 0$. Pick $y \in \bar{K}$ such that $f(x,y) = 0$. Then the pair (x,y) satisfies any equation over K which is satisfied by $(X,\mathfrak{Y})$[†]. Put

$$\bar{c}_{ijk} = r^{(0)}_{ijk}(x) + \dots + r^{(d-1)}_{ijk}(x)y^{d-1}$$

and

$$\bar{g}_i(Z,U) = \sum_{j,k} \bar{c}_{ijk} Z^j U^k \qquad (i = 1,2) .$$

Then $\bar{c}_{ijk} \in \bar{K}$ and

$$g(Z,U) = \bar{g}_1(Z,U)\bar{g}_2(Z,U) ,$$

contradicting the absolute irreducibility of $g(Z,U)$.

This completes the proof of (2.1). The proof of (2.2) is similar but simpler.

<u>LEMMA 2B</u>: <u>Suppose</u> $f(X,Y) \in K[X,Y]$ <u>is of degree</u> $d > 0$ <u>in</u> Y, <u>irreducible over</u> K <u>and separable in</u> Y. <u>Let</u> $f(X,\mathfrak{Y}) = 0$ <u>and</u> $f(Z,\mathfrak{U}) = 0$. <u>Then</u> f <u>is absolutely irreducible if and only if</u>

$$[K(X,Z,\mathfrak{Y},\mathfrak{U}) : K(X,Z)] = d^2 .$$

[†] For if $\ell(X,Y)$ is a polynomial with $\ell(X,\mathfrak{Y}) = 0$, then $\ell(X,Y)$ is divisible by $f(X,Y)$.

Proof: The "only if" part follows from Lemma 2A. The "if" part will be given later[†] in these lectures; we do not need it now.

Let K be a field of characteristic p; let $q = p^{\kappa}$. If

$$f(X,Y) = \sum_{i,j} a_{ij} X^i Y^j \qquad (a_{ij} \in K),$$

define

$$f^{[q]}(X,Y) = \sum_{i,j} a_{ij}^q X^i Y^j .$$

Since the mapping $x \to x^q$ is an automorphism of $\bar{K}$, it follows that if f is absolutely irreducible, then so is $f^{[q]}$.

COROLLARY 2C: *Suppose $f(X,Y)$, $f^{[q]}(X,Y)$ are as above. Suppose f is of degree $d > 0$ in Y. Let X, Z be variables, and let $\mathfrak{Y}, \mathfrak{U}$ be such that*

$$f(X,\mathfrak{Y}) = 0, \qquad f^{[q]}(Z,\mathfrak{U}) = 0 .$$

Then

$$[K(X,Z,\mathfrak{Y},\mathfrak{U}) : K(X,Z)] = d^2 .$$

LEMMA 2D: *Let K be a field of characteristic p. Suppose*

$$f(X,Y) = Y^d + g_1(X)Y^{d-1} + \dots + g_d(X)$$

is a polynomial in $K[X,Y]$, absolutely irreducible and with

$$\deg g_i(X) \le i \qquad (1 \le i \le d) .$$

Let $f(X,\mathfrak{M}) = 0$. If

[†] Ch. V, §3.

$$a(X,Y,Z,W) \neq 0$$

<u>is a polynomial with</u>

$$\text{(i)} \quad \deg_X a \leq (q/d) - d \quad ,$$

$$\text{(ii)} \quad \deg_Y a \leq d - 1 \, ,$$

$$\text{(iii)} \quad \deg_W a \leq d - 1 \, ,$$

<u>then</u>

$$a(X,\mathfrak{Y},X^q,\mathfrak{Y}^q) \neq 0 \quad .$$

Before commencing with the proof, we give some heuristic arguments. Since f is irreducible, the elements

$$\mathfrak{Y}^i \quad (0 \leq i \leq d-1)$$

are linearly independent over $K(X)$. On the other hand, since there are d^2 of them, the elements

$$\mathfrak{Y}^i\mathfrak{Y}^{qk} \quad (0 \leq i , k \leq d-1)$$

are linearly <u>dependent</u> over $K(X)$. Hence the lemma is not trivial. However, the powers of X in $a(X,Y,X^q,Y^q)$ are restricted. We have only the powers

$$X^{qj+v} \quad (0 \leq v \leq (q/d) - d \, ; \, j = 0,1,\ldots) \, .$$

So roughly only one d^{th} of all possible exponents in X can occur. That is why the lemma has a chance of working.

<u>Proof of the lemma</u>: The method is similar to that of Chapter I, §5. Put

$$\hat{a}(X,Y,Z;W_1,\ldots,W_d) = \prod_{i=1}^{d} a(X,Y,Z,W_i) \quad .$$

This is a polynomial in $d+3$ variables, symmetric in $W_1,\ldots,W_d$. By Lemma 5A of Chapter I,

$$\hat{a}(X,Y,Z;W_1,\ldots,W_d) = b(X,Y,Z;s_1(W_1,\ldots,W_d),\ldots,s_d(W_1,\ldots,W_d)) \quad ,$$

where $s_1,\ldots,s_d$ are the elementary symmetric polynomials in $W_1,\ldots,W_d$. By the same lemma, the total degree of $b(X,Y,Z;V_1,\ldots,V_d)$ in $V_1,\ldots,V_d$ is at most $d-1$.

Now since

$$\mathfrak{Y}^d = -g_1(X)\mathfrak{Y}^{d-1} - \ldots - g_d(X) \quad ,$$

we have for any positive integer t,

$$\mathfrak{Y}^{d-1+t} = g_1^{(t)}(X)\mathfrak{Y}^{d-1} + \ldots + g_d^{(t)}(X) \quad , \tag{2.3}$$

where it is easily verified by induction that

$$\deg g_i^{(t)}(X) \le (t-1+i) \qquad (1 \le i \le d) \quad .$$

Since

$$\deg_Y b \le d(d-1) = (d-1) + (d-1)^2 \quad ,$$

we apply (2.3) with $t \le (d-1)^2$, to obtain

$$b(X,\mathfrak{Y},Z\,;V_1,\ldots,V_d) = c(X,\mathfrak{Y},Z\,;V_1,\ldots,V_d) \quad ,$$

where $\deg_Y c \le d-1$. Furthermore,

$$\begin{aligned}\deg_X c &\le \deg_X b + ((d-1)^2 - 1 + d)\\ &= \deg_X b + d(d-1)\\ &\le d \deg_X a + d(d-1)\\ &\le q - d^2 + d(d-1)\\ &< q .\end{aligned}$$

Suppose, indirectly, that

$$a(X,\mathfrak{Y},X^q,\mathfrak{Y}^q) = 0 .$$

Let $\mathfrak{Y}_1 = \mathfrak{Y}$ and $f(X,Y) = (Y - \mathfrak{Y}_1)(Y - \mathfrak{Y}_2) \ldots (Y - \mathfrak{Y}_d)$. Then

$$\hat{a}(X,\mathfrak{Y},X^q ; \mathfrak{Y}_1^q,\ldots,\mathfrak{Y}_d^q) = 0 ;$$

and since

$$s_i(\mathfrak{Y}_1,\ldots,\mathfrak{Y}_d) = g_i(X) , \qquad (1 \le i \le d)$$

whence

$$s_i(\mathfrak{Y}_1^q,\ldots,\mathfrak{Y}_d^q) = g_i^{[q]}(X^q) , \qquad (1 \le i \le d)$$

we have

$$b(X,\mathfrak{Y},X^q ; g_1^{[q]}(X^q),\ldots,g_d^{[q]}(X^q)) = 0 .$$

Therefore

$$c(X,\mathfrak{Y},X^q ; g_1^{[q]}(X^q),\ldots,g_d^{[q]}(X^q)) = 0 .$$

But since $\deg_Y c \le d - 1$, and $\mathfrak{Y}$ is algebraic of degree d , we must have the following identity in two variables:

$$c(X,Y,X^q ; g_1^{[q]}(X^q),\ldots,g_d^{[q]}(X^q)) = 0 .$$

Now make the substitution $X = X_1 + X_2$. Note that $X^q = X_1^q + X_2^q$, so that for some polynomial ℓ,

$$c(X_1+X_2, Y, X_1^q ; g_1^{[q]}(X_1^q), \dots, g_d^{[q]}(X_1^q)) + X_2^q \ell(X_1, X_2, Y) = 0 .$$

Since $\deg_X c < q$, the first summand has a degree strictly smaller than q in X_2, and we obtain the identity

$$c(X_1+X_2, Y, X_1^q ; g_1^{[q]}(X_1^q), \dots, g_d^{[q]}(X_1^q)) = 0 .$$

Since $X_1 + X_2, Y, X_1^q$ are algebraically independent, we may replace them by variables X, Y, Z, to obtain

$$c(X, Y, Z ; g_1^{[q]}(Z), \dots, g_d^{[q]}(Z)) = 0 .$$

Substituting $\mathfrak{Y}$ for Y, we obtain

(2.4) $$b(X, \mathfrak{Y}, Z ; g_1^{[q]}(Z), \dots, g_d^{[q]}(Z)) = 0 .$$

Now let $\mathfrak{U}_1, \dots, \mathfrak{U}_d$ be quantities with

$$f^{[q]}(Z, U) = (U - \mathfrak{U}_1) \dots (U - \mathfrak{U}_d) ,$$

whence

$$s_i(\mathfrak{U}_1, \dots, \mathfrak{U}_d) = g_i^{[q]}(Z) \qquad (1 \le i \le d) .$$

By the construction of the polynomial b, and by (2.4),

$$\hat{a}(X, \mathfrak{Y}, Z ; \mathfrak{U}_1, \dots, \mathfrak{U}_d) = 0 .$$

Hence for some i, $1 \le i \le d$, the quantity $\mathfrak{U} = \mathfrak{U}_i$ satisfies

$$a(X, \mathfrak{Y}, Z, \mathfrak{U}) = 0 .$$

But by Corollary 2C, and since $f(X, \mathfrak{Y}) = 0$, $f^{[q]}(Z, \mathfrak{U}) = 0$, the d^2 elements $\mathfrak{Y}^j \mathfrak{U}^k$ $(0 \le j, k \le d-1)$ are linearly independent over

$K(X,Z)$. Therefore $a(X,Y,Z,W)$ must be identically zero. This is a contradiction, and the lemma is established.

§3. Derivatives.

Let $f(X,Y)$ be the polynomial of Theorem 1A. It is of total degree d, and we may assume it to be separable in Y. Let $f_X(X,Y)$, $f_Y(X,Y)$ denote partial derivatives with respect to X, Y, respectively. As before, let $\mathfrak{Y}$ satisfy $f(X,\mathfrak{Y}) = 0$.

Let D be the operator of differentiation with respect to X in $F_q(X)$. Since $\mathfrak{Y}$ is separable over this field, D may uniquely be extended to a derivation in $F_q(X,\mathfrak{Y})$. In fact, $D(f(X,\mathfrak{Y})) = f_X(X,\mathfrak{Y}) + f_Y(X,\mathfrak{Y})D\mathfrak{Y} = 0$, whence

$$D\mathfrak{Y} = -f_X(X,\mathfrak{Y})/f_Y(X,\mathfrak{Y}) \quad . \tag{3.1}$$

LEMMA 3A: *Suppose* $0 \le \ell \le M$. *If* $a(X,Y)$ *is a polynomial, then*

$$D^{\ell}(f_Y^{2M}(X,\mathfrak{Y})\,a(X,\mathfrak{Y})) = f_Y^{2M-2\ell}(X,\mathfrak{Y})\,a^{(\ell)}(X,\mathfrak{Y}) \;,$$

where $a^{(\ell)}(X,Y)$ *is a polynomial with*

$$\deg a^{(\ell)} \le \deg a + (2d-3)\ell \;.$$

Proof: The proof is by induction on ℓ. If $\ell = 0$, there is nothing to prove. Suppose the lemma holds for ℓ, $0 \le \ell < M$. Then

$$\begin{aligned} D^{\ell+1}(f_Y^{2M}(X,\mathfrak{Y})\,a(X,\mathfrak{Y})) &= D(f_Y^{2M-2\ell}(X,\mathfrak{Y})\,a^{(\ell)}(X,\mathfrak{Y})) \\ &= (2M-2\ell)\,f_Y^{2M-2\ell-1}(X,\mathfrak{Y})\,(f_{YX}(X,\mathfrak{Y}) + f_{YY}(X,\mathfrak{Y})D\mathfrak{Y})\,a^{(\ell)}(X,\mathfrak{Y}) \\ &\quad + f_Y^{2M-2\ell}(X,\mathfrak{Y})\,(a_X^{(\ell)}(X,\mathfrak{Y}) + a_Y^{(\ell)}(X,\mathfrak{Y})\,D\mathfrak{Y}) \quad . \end{aligned}$$

Substituting (3.1), we get

$$f_Y^{2M-2(\ell+1)}(X,\mathfrak{Y})\Big((2M-2\ell)f_{YX}(X,\mathfrak{Y})f_Y(X,\mathfrak{Y}) - f_{YY}(X,\mathfrak{Y})f_X(X,\mathfrak{Y}))a^{\ell}(X,\mathfrak{Y})$$

$$+ a_X^{(\ell)}(X,\mathfrak{Y})f_Y^2(X,\mathfrak{Y}) - a_Y^{(\ell)}(X,\mathfrak{Y})f_X(X,\mathfrak{Y})f_Y(X,\mathfrak{Y})\Big)$$

$$= f_Y^{2M-2(\ell+1)}(X,\mathfrak{Y})a^{(\ell+1)}(X,\mathfrak{Y}) ,$$

say. It is then clear that

$$\deg a^{(\ell+1)} \le \deg a^{(\ell)} + (2d-3) \le \deg a + (2d-3)(\ell+1) .$$

LEMMA 3B: *Let*

$$f(X,Y) = Y^d + g_1(X)Y^{d-1} + \dots + g_d(X) ,$$

where

(3.2) $$\deg g_i(X) \le i \qquad (1 \le i \le d) .$$

Suppose

$$f(X,Y) = (Y-\mathfrak{Y}_1)(Y-\mathfrak{Y}_2)\dots(Y-\mathfrak{Y}_d) .$$

If $a(X,Y_1,\dots,Y_d)$ *is a polynomial symmetric in* $Y_1,\dots,Y_d$, *then*

$$a(X,\mathfrak{Y}_1,\dots,\mathfrak{Y}_d) = b(X) ,$$

where $b(X)$ *is a polynomial with*

$$\deg b \le \text{total} \deg a(X,Y_1,\dots,Y_d) .$$

Proof: Let δ denote the total degree of $a(X,Y_1,\dots,Y_d)$. Then

$$a(X,Y_1,\dots,Y_d) = \sum_{v=0}^{\delta} X^v c_v(Y_1,\dots,Y_d) ,$$

where $c_v(Y_1,\ldots,Y_d)$ is a polynomial of degree $\leqq \delta - v$, symmetric in $Y_1,\ldots,Y_d$. By Lemma 5A, Chapter I,

$$c_v(Y_1,\ldots,Y_d) = h_v(s_1(Y_1,\ldots,Y_d),\ldots,s_d(Y_1,\ldots,Y_d)) .$$

Moreover, by the same lemma, any monomial $s_1^{i_1} s_2^{i_2} \ldots s_d^{i_d}$ occurring in $h_v(s_1,\ldots,s_d)$ has $i_1 + 2i_2 + \ldots + di_d \leq \delta - v$. Hence in

$$c_v(\mathfrak{Y}_1,\ldots,\mathfrak{Y}_d) = h_v(g_1(X),\ldots,g_d(X)) ,$$

every summand $g_1^{i_1}(X) g_2^{i_2}(X) \ldots g_d^{i_d}(X)$ has degree at most

$$i_1 + 2i_2 + \ldots + di_d \leq \delta - v$$

by (3.2). Therefore every summand $X^v c_v(\mathfrak{Y}_1,\ldots,\mathfrak{Y}_d)$ of $a(X,\mathfrak{Y}_1,\ldots,\mathfrak{Y}_d)$ is a polynomial of degree $\leq v + \delta - v = \delta$.

§4. Construction of two algebraic functions.

Let $f(Y) = a_0 Y^d + a_1 Y^{d-1} + \ldots + a_d = a_0(Y - y_1) \ldots (Y - y_d)$; thus $y_1,\ldots,y_d$ are the roots of $f(Y)$. The discriminant Δ of f is

$$\Delta = a_0^{2d-2} \prod_{1 \leqq i < j \leqq d} (y_i - y_j)^2 .$$

It is well known (and may be deduced from Lemma 5A of Chapter I), that Δ is a polynomial of degree $2d - 2$ in the coefficients $a_0,\ldots,a_d$. Moreover, every monomial $a_0^{i_0} a_1^{i_1} \ldots a_d^{i_d}$ occuring in this polynomial has

$$(4.1) \qquad i_1 + 2\,i_2 + \ldots + d\,i_d = d(d - 1) .$$

Now let

$$f(X,Y) = Y^d + g_1(X)Y^{d-1} + \dots + g_d(X) ,$$

with

(4.2) $$\deg g_i(X) \le i \qquad (1 \le i \le d) .$$

Let $\Delta(X)$ be the discriminant of $f(X,Y)$ as a polynomial in Y. Clearly $\Delta(X)$ is a polynomial in X. Moreover, by (4.1), (4.2),

(4.3) $$\deg \Delta(X) \leqq d(d-1) .$$

In what follows, we shall assume that

(4.4) $$d \geqq 2 .$$

We may do so, since Theorem 1A is trivial if $d = 1$.

Let $\mathfrak{S}$ be the set of $x \in F_q$ with $\Delta(x) \neq 0$. Then

(4.5) $$q - d(d-1) \le |\mathfrak{S}| \le q .$$

If $x \in \mathfrak{S}$, the polynomial $f(x,Y)$ has d distinct roots $y_1,\dots,y_d \in \overline{F_q}$. We are, of course, interested in those y's which in fact lie in F_q. Let $\mathfrak{T}_1(x)$ be the set of those y's among $y_1,\dots,y_d$ which lie in F_q. Let $\mathfrak{T}_2(x)$ consist of those y's which are not in F_q. Then for every $x \in \mathfrak{S}$,

$$|\mathfrak{T}_1(x)| + |\mathfrak{T}_2(x)| = d .$$

Define $g_0(X) = 1$ and

$$e_1(X,Y,Y') = Y - Y' ,$$

$$e_2(X,Y,Y') = \sum_{j=1}^{d} g_{d-j}(X)(Y^{j-1} + Y^{j-2}Y' + \dots + Y'^{j-1}) .$$

Then

$$f(X,Y) - f(X,Y') = e_1(X,Y,Y')e_2(X,Y,Y') .$$

If $x \in \mathfrak{S}$ and $y \in \mathfrak{T}_1(x) \cup \mathfrak{T}_2(x)$, then

$$0 = f(x,y) = (f(x,y))^q = f(x,y^q) ,$$

whence

$$0 = f(x,y) - f(x,y^q) = (y - y^q)e_2(x,y,y^q) .$$

If $y \in \mathfrak{T}_1(x)$, then $y \in F_q$, so $y - y^q = 0$; and because y is a simple root of $f(x,Y)$, $e_2(x,y,y^q) \neq 0$. If $y \in \mathfrak{T}_2(x)$, then $y^q \neq y$, hence $e_2(x,y,y^q) = 0$. Hence for $\lambda = 1$ or 2, $\mathfrak{T}_\lambda(x)$ *is the set of* y *with*

$$f(x,y) = 0 \text{ and with } e_\lambda(x,y,y^q) = 0 .$$

Notation: Set $\varepsilon_1 = 1$, $\varepsilon_2 = d - 1$. Then e_λ has total degree ε_λ $(\lambda = 1,2)$.

LEMMA 4A: *Suppose* $\lambda = 1$ *or* 2. *Let* M *be a positive integer with*

$$d|M, \quad M \geq d^2, \quad 2(d-1)(M+8)^2 \leq q .$$

Then there exists a polynomial $a(X,Y)$ *such that*

(i) $a(X,\mathfrak{Y}) \neq 0$,

(ii) *if* $a^{(\ell)}(X,Y)$ *is defined as in Lemma 3A, then*

$$a^{(\ell)}(x,y) = 0 \quad (0 \leq \ell < M)$$

for $x \in \mathfrak{S}$ *and* $y \in \mathfrak{T}_\lambda(x)$,

(iii) $\deg a(X,Y) \leq (\varepsilon_\lambda/d)qM + q(d - 3/2)$.

Proof: The idea of the proof is similar to the ideas used to prove Lemmas 3B and 9C in Chapter I. We try

$$a(X,Y) = \sum_{\substack{j=0 \\ j+k \le K}}^{K} \sum_{k=0}^{d-1} b_{jk}(X,Y) X^{qj} Y^{qk}$$

with

$$b_{jk}(X,Y) = \sum_{i=0}^{d-1} a_{ijk}(X) Y^{i} ,$$

where

$$\deg a_{ijk}(X) \le (q/d) - d - i - j - k ,$$

and

$$K = (\varepsilon_\lambda/d) M + d - 2 .$$

By Lemma 2D, if not all $a_{ijk}(X)$ are zero, then

$$a(X,\mathfrak{Y}) \ne 0 .$$

Since the derivatives of X^q and $\mathfrak{Y}^q$ vanish, it is clear that

$$a^{(\ell)}(X,Y) = \sum_{\substack{j=0 \\ j+k \le K}}^{K} \sum_{k=0}^{d-1} b_{jk}^{(\ell)}(X,Y) X^{qj} Y^{qk} ,$$

where, by Lemma 3A,

$$\deg b_{jk}^{(\ell)} \le \deg b_{jk} + (2d-3)\ell \le (q/d) - d - j - k + (2d-3)\ell .$$

We want $a^{(\ell)}(x,y) = 0$ for $0 \le \ell < M$, $x \in \mathfrak{S}$ and $y \in \mathfrak{T}_\lambda(x)$.

Case 1: $\lambda = 1$. Here $x, y \in F_q$, so $x^q = x$ and $y^q = y$. We need to have the polynomial

$$c^{(\ell)}(X,Y) = \sum_{\substack{j=0 \\ j+k\le K}}^{K} \sum_{k=0}^{d-1} b_{jk}^{(\ell)}(X,Y)X^jY^k$$

vanish for the pairs (x,y) under consideration. Notice that

$$\deg c^{(\ell)} \le (q/d) + (2d-3)\ell - 2 .$$

<u>Case 2</u>: $\lambda = 2$. Here $x \in F_q$, $f(x,y) = 0$ and $e_2(x,y,y^q) = 0$. So $x^q = x$ and

$$0 = e_2(x,y,y^q) = y^{q(d-1)} + y^{q(d-2)}y + \dots + y^{d-1}$$
$$+ g_1(x)(y^{q(d-2)} + \dots + y^{d-2}) + \dots + g_{d-1}(x) .$$

Hence we may express $y^{q(d-1)}$ in terms of $1, y^q, \dots, y^{q(d-2)}$, with coefficients which are polynomials in x,y of degree at most $d-1$. That is, we need that a certain polynomial $c^{(\ell)}(X,Y,Y')$ vanishes for (x,y,y^q), where $c^{(\ell)}(X,Y,Y')$ is of degree at most $d-2$ in Y', and of total degree at most $(q/d) + (2d-3)\ell - 2$ in X,Y.

In both cases, we need that a certain polynomial $c^{(\ell)}(X,Y,Y')$ vanishes at (x,y,y^q), where

$\deg c^{(\ell)}$ in X,Y together is $\le (q/d) + (2d-3)\ell - 2$,

$\deg c^{(\ell)}$ in Y' is $\le \varepsilon_\lambda - 1$.

We know that for a pair (x,y) with $f(x,y) = 0$,

$$y^d = -g_1(x)y^{d-1} - \dots - g_d(x) ,$$

and for positive integers t,

$$y^{d-1+t} = g_1^{(t)}(x)y^{d-1} + \ldots + g_d^{(t)}(x) ,$$

where

$$\deg g_i^{(t)}(X) \le t + i - 1 .$$

(See (2.3)). We may express $y^d, y^{d+1}, \ldots$ in terms of $1, y, \ldots, y^{d-1}$. Hence $c^{(\ell)}(x,y,y^q) = 0$ precisely if a certain polynomial $d^{(\ell)}(x,y,y^q) = 0$, where

$$\deg_X d^{(\ell)} \le (q/d) + (2d-3)\ell - 2 ,$$

$$\deg_Y d^{(\ell)} \le d - 1 ,$$

$$\deg_{Y'} d^{(\ell)} \le \varepsilon_\lambda - 1 .$$

Condition (ii) of the lemma is certainly satisfied if $d^{(\ell)}(X,Y,Y')$ is identically zero for $0 \le \ell < M$.

The number of coefficients of $d^{(\ell)}(X,Y,Y')$ is at most

$$\varepsilon_\lambda d((q/d) + (2d-3)\ell - 1) < \varepsilon_\lambda q + (2d^2 - 3d)\varepsilon_\lambda \ell .$$

The number B of coefficients of all polynomials $d^{(\ell)}(X,Y,Y')$, $0 \le \ell < M$, satisfies

$$B < \varepsilon_\lambda qM + \varepsilon_\lambda \tfrac{1}{2} M^2 (2d^2 - 3d) .$$

These coefficients are linear combinations of the coefficients of the $a_{ijk}(X)$. We obtain a system of linear homogeneous equations in the as yet undetermined coefficients of the polynomials $a_{ijk}(X)$.

The number of coefficients available for a_{ijk} is at least

$$(q/d) - d - i - j - k \ge (q/d) - d - 2(d-1) - j > (q/d) - 3d - j .$$

Summing over j, $0 \le j \le K - k$, the number of available coefficients is at least

$$(q/d)(K-k+1)-3d(K+1)-\tfrac{1}{2}(K-k)(K-k+1) = ((q/d)-3d)(K+1)-\tfrac{1}{2}(K-k)(K-k+1)-(q/d)k.$$

Summing over k, $0 \le k \le d-1$, the number of available coefficients is

$$\geqq (q-3d^2)(K+1) - \tfrac{1}{2}K^2 d - (q/d)\tfrac{1}{2}d(d-1).$$

Summing over i, $0 \le i \le d-1$, we obtain the total number A of available coefficients. This number satisfies

$$\begin{aligned} A &> (q-3d^2)(Kd+d) - \tfrac{1}{2}qd(d-1) - \tfrac{1}{2}K^2d^2 \\ &> (q-3d^2)(\varepsilon_\lambda M + d^2 - d) - \tfrac{1}{2}qd(d-1) - \tfrac{1}{2}(\varepsilon_\lambda M + d^2)^2 \\ &> \varepsilon_\lambda qM + q(\tfrac{1}{2}d^2 - \tfrac{1}{2}d) - \tfrac{1}{2}\varepsilon_\lambda^2 M^2 - 6\varepsilon_\lambda Md^2 - 2\varepsilon_\lambda Md^2 , \end{aligned}$$

since $M \geqq d^2$ by hypothesis. In order that the polynomials $d^{(\ell)}(X,Y,Y')$ vanish, we have to solve a homogeneous system of B linear equations in A variables. In order to get a non-zero solution, it is sufficient that $B < A$. We need that

$$\tfrac{1}{2}\varepsilon_\lambda M^2(2d^2 - 3d + \varepsilon_\lambda) + 8\varepsilon_\lambda Md^2 < \tfrac{1}{2}qd(d-1).$$

Since $\varepsilon_\lambda = 1$ or $d-1$, this inequality certainly holds if

$$\tfrac{1}{2}M^2(d-1)(2d^2-2d-1) + 8Md^2(d-1) < \tfrac{1}{2}qd(d-1).$$

Hence it holds if $M^2(d-1) + 8Md < \tfrac{1}{2}q$. But this is true by (4.4) and by our hypothesis that $2(d-1)(M+8)^2 \leqq q$.

Finally,

$$\begin{aligned} \deg a(X,Y) &\le Kq + (q/d) \\ &= (\varepsilon_\lambda/d)qM + q(d-2+(1/d)) \\ &\le (\varepsilon_\lambda/d)qM + q(d-(3/2)) . \end{aligned}$$

This concludes the proof of Lemma 4A.

Remark: Set

$$c(X,Y) = f_Y^{2M}(X,Y)\, a(X,Y) \quad .$$

Then (i) $c(X,\mathfrak{Y}) \neq 0$,

(ii) if we take derivatives for $0 \le \ell < M$, then

$$D^{\ell} c(X,\mathfrak{Y}) = f_Y^{2M-2\ell}(X,\mathfrak{Y})\, a^{(\ell)}(X,\mathfrak{Y}) \quad .$$

Hence for $x \in \mathfrak{S}$, $y \in \mathfrak{T}_{\lambda}(x)$, we have $D^{\ell} c(x,y) = 0$.

(iii) $\deg c \le (\varepsilon_{\lambda}/d)\, q\, M + q(d - (3/2)) + 2\, Md$.

But if $q > 250\, d^5$, then $2\, Md \le 2\, d\sqrt{q} = \frac{2d}{\sqrt{q}}\, q < \frac{1}{2} q$, so that

$$\deg c \le (\varepsilon_{\lambda}/d)\, q\, M + q(d-1) \quad .$$

5. Construction of two polynomials.

LEMMA 5A: Suppose M satisfies the conditions of Lemma 4A. Let $\lambda = 1$ or 2 be fixed. Then there exists a polynomial $r(X) \neq 0$ with

(i) $D^{\ell} r(x) = 0$ for $x \in \mathfrak{S}$ and $0 \le \ell < M|\mathfrak{T}_{\lambda}(x)|$,

(ii) $\deg r(X) \le \varepsilon_{\lambda}\, q\, M + q\, d(d-1)$.

Proof: We have constructed $c(X,\mathfrak{Y})$ in §4. Set $r(X) = \mathfrak{N}(c(X,\mathfrak{Y}))$, where $\mathfrak{N}$ denotes the norm from the field $F_q(X,\mathfrak{Y})$ to $F_q(X)$. So if $f(X,Y) = (Y - \mathfrak{Y}_1)(Y - \mathfrak{Y}_2) \ldots (Y - \mathfrak{Y}_d)$, then

$$r(X) = \prod_{j=1}^{d} c(X,\mathfrak{Y}_j) \quad .$$

Now

$$(5.1) \quad D^{\ell} r(X) = \sum_{u_1 + \dots + u_d = \ell} \left(\frac{\ell!}{u_1! \dots u_d!}\right) (D^{u_1} c(X, \mathfrak{Y}_1)) \dots (D^{u_d} c(X, \mathfrak{Y}_d)) .$$

The R.H.S. of (5.1) is a symmetric polynomial in $\mathfrak{Y}_1, \dots, \mathfrak{Y}_d$; hence, a polynomial in the elementary symmetric functions of $\mathfrak{Y}_1, \dots, \mathfrak{Y}_d$:

$$D^{\ell} r(X) = k(X, g_1(X), \dots, g_d(X)) .$$

So for $x \in F_q$,

$$D^{\ell} r(x) = k(x, g_1(x), \dots, g_d(x)) .$$

If $x \in \mathfrak{S}$, $f(x,Y)$ has d distinct roots $y_1, \dots, y_d \in \overline{F_q}$, and $s_i(y_1, \dots, y_d) = g_i(x)$. Therefore

$$(5.2) \quad D^{\ell} r(x) = \sum_{u_1 + \dots + u_d = \ell} \left(\frac{\ell!}{u_1! \dots u_d!}\right) (D^{u_1} c(x, y_1)) \dots (D^{u_d} c(x, y_d)) .$$

[A sophisticated reader might say that (5.2) is obtained from (5.1) by the specialization $X, \mathfrak{Y}_1, \dots, \mathfrak{Y}_d \to x, y_1, \dots, y_d$.]

We have

$$\{y_1, \dots, y_d\} = \mathfrak{T}_1(x) \cup \mathfrak{T}_2(x) .$$

Suppose, without loss of generality, that

$$y_1, \dots, y_t \in \mathfrak{T}_{\lambda}(x) ,$$

so that $t = |\mathfrak{T}_{\lambda}(x)|$. Each summand of the R.H.S. of (5.2) has

$$u_1 + u_2 + \dots + u_t \le \ell .$$

Therefore for some integer s , $1 \le s \le t$,

$$u_s \le \frac{\ell}{t} = \frac{\ell}{|\mathfrak{X}_\lambda(x)|} < \frac{M|\mathfrak{X}_\lambda(x)|}{|\mathfrak{X}_\lambda(x)|} = M .$$

By part (ii) of the remark at the end of §4,

$$D^{u_s} c(x, y_s) = 0 ,$$

and each summand of (5.2) has a zero factor. Therefore for every $x \in \mathfrak{S}$,

$$D^\ell r(x) = 0 \qquad (0 \le \ell < M|\mathfrak{X}_\lambda(x)|) .$$

Now

$$r(X) = \prod_{j=1}^{d} c(X, \mathfrak{Y}_j)$$

is a polynomial in $X, \mathfrak{Y}_1, \ldots, \mathfrak{Y}_d$, which is symmetric in $\mathfrak{Y}_1, \ldots, \mathfrak{Y}_d$ and is of total degree at most

$$d((\varepsilon_\lambda/d)qM + q(d-1)) = \varepsilon_\lambda qM + qd(d-1) .$$

Hence by Lemma 3B,

$$\deg r(X) \le \varepsilon_\lambda qM + qd(d-1) .$$

The proof of Lemma 5A is complete.

§6. Proof of the Main Theorem.

For the moment, we consider only the case $q = p$. For then for every $x \in \mathfrak{S}$,

$$M|\mathfrak{X}_\lambda(x)| \leqq dM < q = p ,$$

and we need this in order to use Theorem 1G of Chapter I and to conclude that the polynomials $r_\lambda(X)$ constructed in Lemma 5A have zeros of the desired multiplicity. The general case will be treated in §9.

Set

$$N_\lambda = \sum_{x \in \mathfrak{S}} |\mathfrak{T}_\lambda(x)| \qquad (\lambda = 1,2) .$$

Observe that by (4.5),

$$d(q - d(d-1)) \le N_1 + N_2 = d|\mathfrak{S}| \le dq .$$

Clearly the number of zeros of $r_\lambda(X)$, counted with multiplicities, cannot exceed its degree; hence by Lemma 5A, and by Theorem 1G of Chapter I,

$$MN_\lambda \le \deg r_\lambda(X) ,$$

and

$$N_\lambda \le \frac{\deg r_\lambda(X)}{M} \le \varepsilon_\lambda q + \frac{qd(d-1)}{M} .$$

Now N_1 is the number of zeros $(x,y) \in F_q^2$ with $\Delta(x) \ne 0$ of $f(X,Y)$. In view of (4.3), we have

$$N \le N_1 + d(d-1)d < q + d(d-1)(q/M) + d^3 .$$

Also,

$$\begin{aligned} N \ge N_1 &> qd - d^3 - N_2 \\ &\ge qd - d^3 - (d-1)q - d(d-1)(q/M) \\ &= q - d(d-1)(q/M) - d^3 . \end{aligned}$$

Therefore

(6.1) $$|N - q| < d(d-1)(q/M) + d^3 .$$

This inequality holds for all integers M satisfying the conditions of Lemma 4A. Choose M to be the multiple of d with

$$(q/2d)^{\frac{1}{2}} - 5d < M \le (q/2d)^{\frac{1}{2}} - 4d .$$

Then since $d \geqq 2$,

$$M \leq (q/2d)^{\frac{1}{2}} - 8 \quad ,$$

or

$$(M + 8)^2 \leq q/2d \ ,$$

so certainly

$$2(d-1)(M+8)^2 < q \ .$$

Also,

$$M > \left(\frac{q}{2d}\right)^{\frac{1}{2}}\left(1 - \frac{5\sqrt{2}\ d^{3/2}}{q^{\frac{1}{2}}}\right) > \left(\frac{q}{2d}\right)^{\frac{1}{2}} \cdot \tfrac{1}{2} > d^2 \ ,$$

since $q > 250\,d^5$. The assumption that $q > 250\,d^5$ also guarantees that

$$\frac{5\sqrt{2}\,d^{3/2}}{q^{\frac{1}{2}}} < \frac{1}{3} \ .$$

By making the simple observation that if $0 < x < \frac{1}{3}$, then

$$\frac{1}{1-x} < 1 + \frac{3}{2}x \ ,$$

we obtain

$$\frac{1}{M} < \left(\frac{2d}{q}\right)^{\frac{1}{2}}\left(1 + \frac{3}{2}\,\frac{5\sqrt{2}\ d^{3/2}}{q^{\frac{1}{2}}}\right) \ .$$

Finally by (6.1),

$$|N - q| < \sqrt{2}\ d(d-1)d^{\frac{1}{2}}q^{\frac{1}{2}}\left(1 + \frac{8\sqrt{2}\ d^{3/2}}{q^{\frac{1}{2}}}\right) + d^3$$

$$< \sqrt{2}\ d^{5/2}q^{1/2} - \sqrt{2}\ d^{3/2}q^{1/2} + 16d^4 + d^3$$

$$< \sqrt{2}\ d^{5/2}q^{1/2} \ .$$

But this is the assertion of Theorem 1A.

We still have the restriction that $q = p$. In the next sections we shall define hyperderivatives in function fields in order to remove this restriction.

§7. Valuations.

Let K be any field. As usual, K^* is the multiplicative group of K.

Definition: A valuation is a mapping v from K^* onto the ring $\mathbb{Z}$, of integers such that

(i) $v(ab) = v(a) + v(b)$,

(ii) $v(a+b) \geq \min\{v(a), v(b)\}$,

with the additional convention that $v(0) = +\infty$.

Let K_0 be the set of $a \in K$ with $v(a) \geq 0$. It is easy to see that K_0 is a subring of K, and that the units of K_0 are precisely the elements $a \in K_0$ with $v(a) = 0$.

Let K_1 be the set of $a \in K$ with $v(a) \geq 1$. It is clear that $K_1 \subseteq K_0$, and that K_1 is closed under addition and subtraction. In fact, K_1 is an ideal in K_0, since if $a \in K_0$, $b \in K_1$, then

$$v(ab) = v(a) + v(b) \geq 0 + 1 = 1,$$

so that $ab \in K_1$. Moreover, any proper ideal in K_0 must not contain a unit, so must not contain any element a with $v(a) = 0$, hence must be contained in K_1. That is, K_1 is a maximal ideal in K_0; in fact, K_1 is the unique maximal ideal in K_0. We summarize in

LEMMA 7A: Let v be a valuation of a field K. Let K_0 be the set of $a \in K$ with $v(a) \geq 0$, and K_1 the set of $a \in K$ with $v(a) \geq 1$. Then K_0 is a subring of K, and K_1 is the unique maximal ideal in K_0. Hence K_0/K_1 is a field.

Example: Let $K = \mathbb{Q}$, and p any prime. Any non-zero rational number can be written in the form $(a/b)p^\nu$, $p \nmid ab$, where ν is unique. Put

$$v((a/b)p^\nu) = \nu .$$

Then it is easy to check that v is a valuation. Now $\mathbb{Q}_0$ is the ring consisting of zero and of elements $(a/b)p^\nu$ with $\nu \geq 0$, and $\mathbb{Q}_1$ is the unique maximal ideal in $\mathbb{Q}_0$, consisting of zero and of elements $(a/b)p^\nu$ with $\nu \geq 1$. A complete set of representatives of $\mathbb{Q}_0$ modulo $\mathbb{Q}_1$ is $\{0,1,2,\ldots,p-1\}$. For if $(a/b)p^\nu \in \mathbb{Q}_0$, pick the integer x in $\{0,1,\ldots,p-1\}$ with

$$ap^\nu \equiv bx \pmod{p} .$$

Then $\frac{a}{b}p^\nu - x = \frac{ap^\nu - bx}{b} \in \mathbb{Q}_1$, so that x lies in the same coset modulo $\mathbb{Q}_1$ as $(a/b)p^\nu$. It follows that $\mathbb{Q}_0/\mathbb{Q}_1$ is a field with p elements, whence

$$\mathbb{Q}_0/\mathbb{Q}_1 \cong F_p .$$

LEMMA 7B: Suppose K is a field with a valuation v, and φ is a homomorphism from K_0 onto a field F with kernel K_1. Let X be a variable. Then there exists an extension v' of v to

$K(X)$ with $v'(X) = 0$, and an extension φ' of φ where $\varphi' \colon (K(X))_0 \to F(X)$, such that $\varphi'(X) = X$, φ' is onto, and the kernel of φ' is $(K(X))_1$.

Proof: First define φ' on $K_0[X]$ by

$$\varphi'(a_0 + a_1 X + \cdots + a_t X^t) = \varphi(a_0) + \varphi(a_1)X + \cdots + \varphi(a_t)X^t .$$

It is clear that φ' is a homomorphism and that φ' extends φ.

Next, define v' on $K[X]$ by

$$v'(a_0 + a_1 X + \cdots + a_t X^t) = \min(v(a_0),\ldots,v(a_t)) .$$

Clearly,

$$v'(f(X) + g(X)) \geq \min(v'(f(X)),v'(g(X))) .$$

We claim that

$$v'(f(X)g(X)) = v'(f(X)) + v'(g(X)) . \tag{7.1}$$

There exists an element $p \in K$ with $v(p) = 1$, since v is onto. Put

$$\hat{f}(X) = p^{-v'(f)} f(X)$$
$$\hat{g}(X) = p^{-v'(g)} g(X) .$$

Then $v'(\hat{f}) = v'(\hat{g}) = 0$, and it suffices to show that $v'(\hat{f}\hat{g}) = 0$, since then

$$v'(fg) = v'(f) + v'(g) + v'(\hat{f}\hat{g}) = v'(f) + v'(g) .$$

We may therefore assume without loss of generality that $v'(f) = v'(g) = 0$. We wish to show $v'(fg) = 0$. But since $v'(f) = 0$, $f(X) \in K_0[X]$, and similarly $g(X) \in K_0[X]$; therefore $f(X)g(X) \in K_0[X]$, and $v'(fg) \geq 0$. Suppose we had $v'(fg) \geq 1$. Then $f(X)g(X) \in K_1[X]$ and

$$\varphi'(f)\varphi'(g) = \varphi'(fg) = 0 .$$

So either $\varphi'(f) = 0$ or $\varphi'(g) = 0$, hence either $v'(f) \geq 1$ or $v'(g) \geq 1$, which is a contradiction. Therefore $v'(fg) = 0$. The proof of (7.1) is complete.

Hence if in general v' is defined by

$$v'\left(\frac{f(X)}{g(X)}\right) = v'(f(X)) - v'(g(X)) ,$$

then v' becomes a valuation of $K(X)$.

To further extend φ , notice that every element of $(K(X))_0$ is of the form $(f(X)/g(X))$, where $v'(f) \geq 0$ and $v'(g) = 0$. (If necessary, multiply both f and g by a suitable power of $p \in K$, where $v(p) = 1$) . Define φ' on $(K(X))_0$ by

$$\varphi'\left(\frac{f(X)}{g(X)}\right) = \frac{\varphi'(f(X))}{\varphi'(g(X))} .$$

It is easy to check that φ' is a well-defined homomorphism from $(K(X))_0$ onto $F(X)$ with kernel $(K(X))_1$.

<u>Example</u>: Let $K = \mathbb{Q}$. Write every non-zero rational as $\frac{a}{b} 3^{\nu}$ where $3 \nmid ab$, and define

$$v\left(\frac{a}{b} 3^{\nu}\right) = \nu .$$

Then, for example,

$$v'\left(\frac{5X + 6}{X^2 + 4}\right) = 0 - 0 = 0 , \qquad \varphi'\left(\frac{5X + 6}{X^2 + 4}\right) = \frac{2X}{X^2 + 1} .$$

LEMMA 7C: *Let* v *be a valuation of a field* K. *Let* φ *be a homomorphism of* K_0 *onto* F, *with kernel* K_1. *Let* η *be algebraic over* F. *Then there exists an element* $\hat{\eta}$ *which is algebraic over* K, *such that* $\hat{\eta}$ *is separable over* K *if* η *is separable over* F. *There exists a valuation* v'' *of* $K(\hat{\eta})$, *with* $v''(\hat{\eta}) = 0$, *extending* v; *and there is a homomorphism* φ'' *of* $K(\hat{\eta})_0$ *onto* $F(\eta)$ *extending* φ, *such that the kernel of* φ'' *is* $K(\hat{\eta})_1$.

$$\begin{array}{ccc} K & \subseteq & K(\hat{\eta}) \\ v\downarrow & & \downarrow v'' \\ \mathbb{Z}\cup\{\infty\} & = & \mathbb{Z}\cup\{\infty\} \end{array} \qquad \begin{array}{ccc} K_0 & \subseteq & K(\hat{\eta})_0 \\ \varphi\downarrow & & \downarrow \varphi'' \\ F & \subsetneq & F(\eta) \end{array}$$

Proof: Let $f(X)$ be the irreducible defining polynomial of η over F. We may choose $f(X)$ to have leading coefficient 1. Let $\hat{f}(X)$ be a polynomial in $K_0[X]$ with the same degree as f, leading coefficient 1, and with $\varphi'(\hat{f}(X)) = f(X)$, where φ' is the epimorphism constructed in Lemma 7B.

We claim that $\hat{f}(X)$ *is irreducible over* K. Suppose, by way of contradiction, that $\hat{f}(X) = \hat{f}_1(X)\hat{f}_2(X)$ is a proper factorization. We may assume that $v'(\hat{f}_1) \geq 0$ and $v'(\hat{f}_2) \geq 0$. (Otherwise, multiply by appropriate powers of an element p of K with $v(p) = 1$.) Then $\hat{f}_1, \hat{f}_2 \in K_0[X]$, and

$$f = \varphi'(\hat{f}) = \varphi'(\hat{f}_1)\varphi'(\hat{f}_2) = f_1 f_2$$

provides a proper factorization of f, which gives a contradiction.

Pick a root, say $\hat{\eta}$, of $\hat{f}(X)$. It is clear that if η is separable over F, then $\hat{\eta}$ is separable over K.

Now define φ'' on $K_0[\hat{\eta}]$ by

$$\varphi''(a_0 + a_1\hat{\eta} + \dots + a_t\hat{\eta}^t) = \varphi(a_0) + \varphi(a_1)\eta + \dots + \varphi(a_t)\eta^t .$$

φ'' is a homomorphism onto $F[\eta] = F(\eta)$. Also define v'' on $K(\hat{\eta})$ by

$$v''(a_0 + a_1\hat{\eta} + \dots + a_{d-1}\hat{\eta}^{d-1}) = \min\{v(a_0), v(a_1), \dots, v(a_{d-1})\} ,$$

where

$$d = \text{degree of } \eta \text{ over } F = \text{degree of } \hat{\eta} \text{ over } K .$$

It is easily verified that v'' is a valuation of $K(\hat{\eta})$, extending v . The proof that for $\alpha, \beta \in K(\hat{\eta})$,

$$v''(\alpha\beta) = v''(\alpha) + v''(\beta) ,$$

goes as the proof of (7.1) in Lemma 7B. The rest of Lemma 7C now follows after noting that $K(\hat{\eta})_0 = K_0[\hat{\eta}]$.

Example: Let $K = \mathbb{Q}$, and p a prime. We define as before,

$$v\left(\frac{a}{b}\, p^{\nu}\right) = \nu \quad \text{if} \quad p \nmid ab .$$

We have seen that there is a homomorphism φ from $\mathbb{Q}_0$ onto F_p with kernel $\mathbb{Q}_1$. The field F_q where $q = p^{\kappa}$, is of the type $F_q = F_p(\eta)$, with η separable algebraic of degree κ . Let $\hat{\eta}$ be chosen as in the lemma and write $N = \mathbb{Q}(\hat{\eta})$. Then there is a valuation v'' of the field $N = \mathbb{Q}(\hat{\eta})$ extending v . Also there is a homomorphism φ'' from N_0 onto F_q with kernel N_1 .

$$\begin{array}{ccc} \mathbb{Q} & \subseteq & N = \mathbb{Q}(\hat{\eta}) \\ v \downarrow & & \downarrow v'' \\ \mathbb{Z} \cup \{\infty\} & = & \mathbb{Z} \cup \{\infty\} \end{array} \qquad \begin{array}{ccc} \mathbb{Q}_0 & \subseteq & N_0 \\ \varphi \downarrow & & \downarrow \varphi'' \\ F_p & \subseteq & F_q \end{array}$$

Remark. It is clear that N is a number field of degree κ. Also, experts in algebraic number theory will say that p is "inertial" in N.

The assertions of the following exercises will not be needed in the sequel.

Exercise 1. Show that every field of characteristic $p \neq 0$ is the homomorphic image of an integral domain of characteristic 0. (For general fields, an appeal to Zorn's Lemma is necessary. It is not necessary for fields which are finitely generated over F_p).

Exercise 2. Let v be a valuation of a field K. Given a monic polynomial $f(Y) = Y^d + a_1 Y^{d-1} + \dots + a_d$ with coefficients in K, put $\psi(f) = \min_{1 \leq i \leq d} (1/i) v(a_i)$. Show that for monic polynomials f, g, we have $\psi(fg) = \min(\psi(f), \psi(g))$. Deduce that if $\deg f = d$ and $\psi(f) = m/d$ with $(m,d) = 1$, then f is irreducible. (If $K = F(X)$ and if $v(a(X)/b(X)) = \deg b(X) - \deg a(X)$, these results reduce to Theorem 1B, Lemma 1C. If $K = \mathbb{Q}$ and if $v((a/b)p^{\nu}) = \nu$, our irreducibility criterion yields Eisenstein's criterion.)

§8. Hyperderivatives again.

In §6 of Chapter I we defined hyperderivatives for polynomials. In the present section we shall more generally define hyperderivatives for algebraic functions. For another approach to hyperderivatives (Hasse derivatives) see Hasse (1936 a), Teichmüller (1936).

Let F_q be a finite field of characteristic p. We have a valuation of $\mathbb{Q}$ given by

$$v\left(\frac{a}{b}\, p^{\nu}\right) = \nu \quad \text{if } p \nmid ab .$$

Associated with this valuation v of $\mathbb{Q}$ is a homomorphism φ from $\mathbb{Q}_0$ onto F_p with kernel $\mathbb{Q}_1$. We can then by Lemma 7C find a field $N \supseteq \mathbb{Q}$ such that v can be extended to a valuation v' of N, and by Lemma 7B further extended to a valuation v'' of $N(X)$. Moreover,

φ can be extended to a homomorphism φ' from N_0 onto F_q with kernel N_1, and φ' can be extended to a homomorphism φ'' from $N(X)_0$ onto $F_q(X)$ with kernel $N(X)_1$. Suppose $f(X,Y) \in F_q[X,Y]$ is an irreducible polynomial which is separable in Y. Let $\mathfrak{Y}$ be an algebraic function with $f(X,\mathfrak{Y}) = 0$. Then there is by Lemma 7C an element $\hat{\mathfrak{Y}}$ which is separable algebraic over $N(X)$, such that we may extend v'' to a valuation v''' of $N(X,\hat{\mathfrak{Y}})$, and φ'' to a homomorphism φ''' from $N(X,\hat{\mathfrak{Y}})_0$ onto $F_q(X,\mathfrak{Y})$ having kernel $N(X,\hat{\mathfrak{Y}})_1$.

$$\begin{array}{ccccccc} Q_0 & \subseteq & N_0 & \subseteq & (N(X))_0 & \subseteq & (N(X,\hat{\mathfrak{Y}}))_0 \\ \downarrow \varphi & & \downarrow \varphi' & & \downarrow \varphi'' & & \downarrow \varphi''' \\ F_p & \subseteq & F_q & \subseteq & F_q(X) & \subseteq & F_q(X,\mathfrak{Y}) \end{array}$$

Hereafter, v, v', v'', v''' are all denoted by v, and $\varphi, \varphi', \varphi'', \varphi'''$ are all denoted by φ. Elements in fields of characteristic zero will be written as $\hat{\mathfrak{Y}}, \hat{u}, \hat{a}(X)$, etc.

Let D be the differentiation operator on $N(X)$. D may be extended to a derivation on $N(X,\hat{\mathfrak{Y}})$, since the extension $N(X,\hat{\mathfrak{Y}})$ over $N(X)$ is separable. We introduce an operator $E^{(\ell)}$ on $N(X,\hat{\mathfrak{Y}})$ by

$$E^{(\ell)}(\hat{u}) = \frac{1}{\ell!} D^{\ell}(\hat{u}) .$$

One verifies immediately that

$$(8.1) \quad E^{(\ell)}(\hat{u}_1 \dots \hat{u}_t) = \sum_{u_1 + \dots + u_t = \ell} (E^{(u_1)}(\hat{u}_1)) \dots (E^{(u_t)}(\hat{u}_t)) .$$

LEMMA 8A: For any $\hat{\mathfrak{u}} \in N(X,\hat{\mathfrak{Y}})$,

$$v(E^{(\ell)}(\hat{\mathfrak{u}})) \geq v(\hat{\mathfrak{u}}) . \quad (\ell = 0,1,2,\ldots)$$

Proof: The proof is by induction on ℓ. The case $\ell = 0$ is trivial. To go from $\ell - 1$ to ℓ, we consider three cases.

(i) The lemma is obvious if $\hat{\mathfrak{u}} \in N[X]$.

(ii) Suppose $\hat{\mathfrak{u}} \in N(X)$. Let $\hat{\mathfrak{u}} = \hat{f}(X)/\hat{g}(X)$, so that $\hat{f}(X) = \hat{g}(X)\hat{\mathfrak{u}}$. By (8.1),

$$E^{(\ell)}(\hat{f}(X)) = \sum_{j=0}^{\ell} (E^{(\ell-j)}\hat{g}(X))E^{(j)}\hat{\mathfrak{u}} . \tag{8.2}$$

Since $\hat{f}(X),\hat{g}(X) \in N[X]$ and by induction on ℓ, the left hand side of (8.2) and every summand on the right hand side of (8.2), except possibly the summand $\hat{g}(X)E^{(\ell)}\hat{\mathfrak{u}}$, has a valuation $\geqq v(\hat{f}(X)) = v(\hat{g}(X)) + v(\hat{\mathfrak{u}})$. Hence also $v(\hat{g}(X)E^{(\ell)}\hat{\mathfrak{u}}) \geqq v(\hat{g}(X)) + v(\hat{\mathfrak{u}})$, which yields $v(E^{(\ell)}\hat{\mathfrak{u}}) \geqq v(\hat{\mathfrak{u}})$.

(iii) Any $\hat{\mathfrak{u}} \in N(X,\hat{\mathfrak{Y}})$ may be written as

$$\hat{\mathfrak{u}} = \hat{r}_0(X) + \hat{r}_1(X)\hat{\mathfrak{Y}} + \ldots + \hat{r}_{d-1}(X)\hat{\mathfrak{Y}}^{d-1}$$

with $\hat{r}_0(X),\hat{r}_1(X),\ldots,\hat{r}_{d-1}(X) \in N(X)$. Since

$$v(\hat{\mathfrak{u}}) = \min\{v(\hat{r}_0(X)),\ldots,v(\hat{r}_{d-1}(X))\} ,$$

it suffices to show that for $0 \leq i \leq d - 1$,

$$v(E^{(\ell)}(\hat{r}_i(X)\hat{\mathfrak{Y}}^i)) \geq v(\hat{r}_i(X)\hat{\mathfrak{Y}}^i) = v(\hat{r}_i(X)) .$$

Applying (8.1) to the product $\hat{r}_i(X)\hat{\mathfrak{Y}}^i = \hat{r}_i(X)\hat{\mathfrak{Y}} \ldots \hat{\mathfrak{Y}}$, it becomes clear that we need only show that

$$v(E^{(\ell)}(\hat{\mathfrak{Y}})) \geq 0 .$$

Let

$$f(X,Y) = Y^d + g_1(X)Y^{d-1} + \dots + g_d(X) .$$

Now $\hat{\mathfrak{Y}}$ was constructed as the root of a polynomial

$$\hat{f}(X,Y) = Y^d + \hat{g}_1(X)Y^{d-1} + \dots + \hat{g}_d(X) ,$$

where $\varphi(\hat{g}_i(X)) = g_i(X)$ $(1 \leqq i \leqq d)$. We have

$$(8.3) \qquad 0 = \hat{f}(X,\hat{\mathfrak{Y}}) = \sum_{i=0}^{d} \hat{g}_{d-i}(X)\hat{\mathfrak{Y}}^i ,$$

and by (8.1),

$$E^{(\ell)}(\hat{g}_{d-i}(X)\hat{\mathfrak{Y}}^i) = E^{(\ell)}(\hat{g}_{d-i}(X)\hat{\mathfrak{Y}} \dots \hat{\mathfrak{Y}})$$

$$= \sum_{u_0+\dots+u_i=\ell} E^{(u_0)}(\hat{g}_{d-i}(X))E^{(u_1)}(\hat{\mathfrak{Y}}) \dots E^{(u_i)}(\hat{\mathfrak{Y}})$$

$$= \sum_{u_0+\dots+u_i=\ell} \hat{\mathfrak{S}}(u_0,\dots,u_i) ,$$

say. Collecting the terms where one of $u_1,\dots,u_i$ equals ℓ , we obtain

$$i\hat{g}_{d-i}(X)\hat{\mathfrak{Y}}^{i-1}E^{(\ell)}(\hat{\mathfrak{Y}}) + \sum_{\substack{u_0+\dots+u_i=\ell \\ u_1,u_2,\dots,u_i<\ell}} \hat{\mathfrak{S}}(u_0,u_1,\dots,u_i) .$$

Hence by (8.3),

$$0 = E^{(\ell)}(\hat{\mathfrak{Y}})\hat{f}_Y(X,\hat{\mathfrak{Y}}) + \sum_{i=0}^{d} \sum_{\substack{u_0+\dots+u_i=\ell \\ u_1,\dots,u_i<\ell}} \hat{\mathfrak{S}}(u_0,\dots,u_i) .$$

But by induction hypothesis, every summand, except possibly the first one, has a valuation $\geqq 0$. Hence also the first one has, i.e.,

$$v(E^{(\ell)}(\hat{\mathfrak{Y}})) + v(\hat{f}_Y(X,\hat{\mathfrak{Y}})) \geq 0 .$$

Since $\hat{f}$ has coefficients in N_0 ,

$$v(\hat{f}_Y(X,\hat{\mathfrak{Y}})) \geq 0 .$$

But $\varphi(\hat{f}_Y(X,\hat{\mathfrak{Y}})) = f_Y(X,\mathfrak{Y}) \neq 0$, and hence

$$v(\hat{f}_Y(X,\hat{\mathfrak{Y}})) = 0 , \tag{8.4}$$

since otherwise $\hat{f}_Y(X,\hat{\mathfrak{Y}}) \in N(X,\hat{\mathfrak{Y}})_1$ (= kernel of φ), a contradiction. It follows that

$$v(E^{(\ell)}(\hat{\mathfrak{Y}})) \geq 0 ,$$

and the proof of the lemma is complete.

We are going to define operators $E^{(\ell)}$ on $F_q(X,\mathfrak{Y})$. Suppose $\mathfrak{u} \in F_q(X,\mathfrak{Y})$. Then there exist $\hat{\mathfrak{u}} \in N(X,\hat{\mathfrak{Y}})_0$ with $\varphi(\hat{\mathfrak{u}}) = \mathfrak{u}$. By Lemma 8A,

$$v(E^{(\ell)}(\hat{\mathfrak{u}})) \geq v(\hat{\mathfrak{u}}) \geq 0 ,$$

whence $E^{(\ell)}(\hat{\mathfrak{u}}) \in N(X,\hat{\mathfrak{Y}})_0$. Define $E^{(\ell)}$ on $F_q(X,\mathfrak{Y})$ by

$$E^{(\ell)}(\mathfrak{u}) = \varphi(E^{(\ell)}(\hat{\mathfrak{u}})) .$$

The new operators $E^{(\ell)}$ are well-defined, because if $\varphi(\hat{\mathfrak{u}}_1) = \varphi(\hat{\mathfrak{u}}_2) = \mathfrak{u}$, then $\varphi(\hat{\mathfrak{u}}_1 - \hat{\mathfrak{u}}_2) = 0$, whence

$$v(E^{(\ell)}(\hat{\mathfrak{u}}_1) - E^{(\ell)}(\hat{\mathfrak{u}}_2)) = v(E^{(\ell)}(\hat{\mathfrak{u}}_1 - \hat{\mathfrak{u}}_2)) \geq v(\hat{\mathfrak{u}}_1 - \hat{\mathfrak{u}}_2) \geq 1 ,$$

so that $\varphi(E^{(\ell)}(\hat{\mathfrak{u}}_1) - E^{(\ell)}(\hat{\mathfrak{u}}_2)) = 0$, whence

$$\varphi(E^{(\ell)}(\hat{\mathfrak{U}}_1)) = \varphi(E^{(\ell)}(\hat{\mathfrak{U}}_2)) .$$

An immediate consequence of our definition and the formula (8.1) for $E^{(\ell)}$ in $N(X,\hat{\mathfrak{Y}})$ is

$$(8.5)\quad E^{(\ell)}(\mathfrak{U}_1\ldots\mathfrak{U}_t) = \sum_{u_1+\ldots+u_t=\ell} (E^{(u_1)}(\mathfrak{U}_1))\ldots(E^{(u_t)}(\mathfrak{U}_t)) .$$

<u>Remark:</u> In the definition of the operators $E^{(\ell)}$ on $F_q(X,\mathfrak{Y})$, we constructed the field $N(X,\hat{\mathfrak{Y}})$, which is not uniquely determined by $F_q(X,\mathfrak{Y})$. Conceivably, the operators $E^{(\ell)}$ could depend on this construction. In fact, the operators $E^{(\ell)}$ are independent of the construction.

A sketch of the proof is as follows. We proceed by induction on ℓ. In the step from $\ell-1$ to ℓ we consider three cases, which are analogous to those in the proof of Lemma 8A.

(i) $\mathfrak{U} \in F_q[X]$. In this case it is easily seen that our hyperderivatives coincide with those defined in §6 of Chapter I. Incidentally, we note for later that Theorem 6D of Chapter I is valid.

(ii) $\mathfrak{U} \in F_q(X)$. Say $\mathfrak{U} = f(X)/g(X)$. By (8.5) and in complete analogy with (8.2),

$$E^{(\ell)}f(X) = \sum_{j=0}^{\ell} (E^{(\ell-j)}g(X))E^{(j)}\mathfrak{U} .$$

Since $f(X),g(X) \in F_q[X]$, and by induction on ℓ, the left hand side and every summand on the right hand side, except possibly the summand $g(X)E^{(\ell)}\mathfrak{U}$, is independent of our construction. Hence also this summand, whence also $E^{(\ell)}\mathfrak{U}$, is independent of our construction.

(iii) $\mathfrak{U} \in F_q(X,\mathfrak{Y})$. The argument is analogous to that in part (iii) of Lemma 8A.

LEMMA 8B: Let $\mathfrak{U} \in F_q(X,\mathfrak{Y})$. Suppose $0 < \ell < p^{\mu}$. Then

$$E^{(\ell)}(\mathfrak{U}^{p^{\mu}}) = 0 .$$

Proof: Pick $\hat{\mathfrak{U}} \in N(X,\hat{\mathfrak{Y}})_0$ with $\varphi(\hat{\mathfrak{U}}) = \mathfrak{U}$. Then

$$E^{(\ell)}(\hat{\mathfrak{U}}^{p^{\mu}}) = \frac{1}{\ell!} D^{(\ell)}(\hat{\mathfrak{U}}^{p^{\mu}}) = \frac{p^{\mu}}{\ell}\left(\frac{1}{(\ell-1)!} D^{(\ell-1)}(\hat{\mathfrak{U}}^{p^{\mu}-1} D\hat{\mathfrak{U}})\right) .$$

We have

$$v\left(\frac{1}{(\ell-1)!} D^{(\ell-1)}(\hat{\mathfrak{U}}^{p^{\mu}-1} D\hat{\mathfrak{U}})\right) = v(E^{(\ell-1)}(\hat{\mathfrak{U}}^{p^{\mu}-1} D\hat{\mathfrak{U}}))$$

$$\geq v(\hat{\mathfrak{U}}^{p^{\mu}-1} D\hat{\mathfrak{U}}) \geq 0 .$$

Since $0 < \ell < p^{\mu}$, $v\left(\frac{p^{\mu}}{\ell}\right) > 0$. Therefore $v(E^{(\ell)}\hat{\mathfrak{U}}^{p^{\mu}}) > 0$, so that

$$E^{(\ell)}(\mathfrak{U}^{p^{\mu}}) = \varphi(E^{(\ell)}\hat{\mathfrak{U}}^{p^{\mu}}) = 0 .$$

§9. Removal of the condition that $q = p$.

We prove the analogue to Lemma 3A:

LEMMA 9A: Let $f(X,Y)$ and $\mathfrak{Y}$ be given as usual. Let M be a positive integer and $a(X,Y)$ a polynomial. Then for $0 \leq \ell \leq M$,

$$(9.1) \qquad E^{(\ell)}(f_Y^{2M}(X,\mathfrak{Y})\, a(X,\mathfrak{Y})) = f_Y^{2M-2\ell}(X,\mathfrak{Y})\, a^{(\ell)}(X,\mathfrak{Y}) ,$$

where $a^{(\ell)}(X,Y)$ is a polynomial with

$$\deg a^{(\ell)}(X,Y) \leq \deg a(X,Y) + (2d-3)\ell .$$

Proof: Find a polynomial $\hat{a}(X,Y)$ in $N_0[X,Y]$, of the same degree as $a(X,Y)$, with $\varphi(\hat{a}(X,Y)) = a(X,Y)$ [†]. Lemma 3A did not depend on the ground field F_q. If we apply this lemma to $D^{\ell}(\hat{f}_Y^{2M}(X,\hat{\mathfrak{Y}})\hat{a}(X,\hat{\mathfrak{Y}}))$ and divide by $\ell!$, we obtain

$$(9.2) \qquad E^{(\ell)}(\hat{f}_Y^{2M}(X,\hat{\mathfrak{Y}})\hat{a}(X,\hat{\mathfrak{Y}})) = \hat{f}_Y^{2M-2\ell}(X,\hat{\mathfrak{Y}})\hat{a}^{(\ell)}(X,\hat{\mathfrak{Y}}) ,$$

where

$$\deg \hat{a}^{(\ell)}(X,Y) \le \deg \hat{a}(X,Y) + (2d-3)\ell .$$

We may suppose that $\hat{a}^{(\ell)}(X,Y)$ is of degree at most $d-1$ in Y, because we may use the relation $\hat{f}(X,\hat{\mathfrak{Y}}) = 0$ to express $\hat{\mathfrak{Y}}^d, \hat{\mathfrak{Y}}^{d+1}, \ldots$, etc., as linear combinations of $1, \hat{\mathfrak{Y}}, \ldots, \hat{\mathfrak{Y}}^{d-1}$. This process does not increase the total degree of the polynomial.

We have

$$v(\hat{f}_Y^{2M-2\ell}(X,\hat{\mathfrak{Y}})\hat{a}^{(\ell)}(X,\hat{\mathfrak{Y}})) \ge 0 ,$$

but $v(\hat{f}_Y(X,\hat{\mathfrak{Y}})) = 0$ by (8.4), whence $v(\hat{a}^{(\ell)}(X,\hat{\mathfrak{Y}})) \ge 0$. Let

$$\hat{a}^{(\ell)}(X,\hat{\mathfrak{Y}}) = \hat{b}_0(X) + \hat{b}_1(X)\hat{\mathfrak{Y}} + \cdots + \hat{b}_{d-1}(X)\hat{\mathfrak{Y}}^{d-1} ;$$

then by our definition of v on $N(X,\hat{\mathfrak{Y}})$,

$$v(\hat{b}_i(X)) \ge 0 \qquad (0 \le i \le d-1) .$$

Thus $\hat{a}^{(\ell)}(X,Y)$ lies in $N_0[X,Y]$. We may therefore apply φ to $\hat{a}^{(\ell)}(X,Y)$; let

$$a^{(\ell)}(X,Y) = \varphi(\hat{a}^{(\ell)}(X,Y)) .$$

Applying φ to (9.2), we obtain (9.1).

We wish to prove the analogue of Lemma 4A, where the higher derivatives D^{ℓ} are replaced by the operators $E^{(\ell)}$. We set

---

[†] Clearly φ may be extended not only to $N_0[X]$, but also in an obvious way to $N_0[X,Y]$.

$$h(X,Y,Z,W) = \sum_{\substack{j=0 \\ j+k\leqq K}}^{K} \sum_{k=0}^{d-1} b_{jk}(X,Y)Z^jW^k ,$$

and put $a(X,Y) = h(X,Y,X^q,Y^q)$. We are interested in

$$E^{(\ell)}(f_Y^{2M}(X,\mathfrak{Y})a(X,\mathfrak{Y})) = f_Y^{2M-2\ell}(X,\mathfrak{Y})a^{(\ell)}(X,\mathfrak{Y}) .$$

But

$$a^{(\ell)}(X,\mathfrak{Y}) = \sum_{\substack{j=0 \\ j+k\leqq K}}^{K} \sum_{k=0}^{d-1} b_{jk}^{(\ell)}(X,\mathfrak{Y})X^{qj}\mathfrak{Y}^{qk} ;$$

this follows from (8.5) and the fact that if $m < M \leq q = p^{\kappa}$, then by Lemma 8B,

$$E^{(m)}(X^{qj}) = 0, \quad E^{(m)}(\mathfrak{Y}^{qk}) = 0 .$$

The remainder of the proof is exactly the same as the proof of Lemma 4A. In this way we obtain an analogue to Lemma 4A.

The rest of the proof of Theorem 1A in the general case is carried out exactly as in the special case $q = p$. No further difficulties arise. But of course we have to use Theorem 6D of Chapter I instead of Theorem 1G of Chapter I.

IV. Equations in Many Variables

References: Chevalley (1935), Warning (1935), Weil (1949), Borevich & Shafarevich (1966), Ax (1964), Joly (1973).

§1. Theorems of Chevalley and Warning.

We adopt the notation $\underline{\underline{X}} = (X_1, X_2, \ldots, X_n)$ for an n-tuple of variables, and $\underline{\underline{x}} = (x_1, x_2, \ldots, x_n)$ for an n-tuple in F_q^n or $\bar{F}_q^n$, i.e. a point of F_q^n or $\bar{F}_q^n$.

LEMMA 1A: *Suppose* u *is an integer with* $0 \le u < q-1$. *Then*

$$\sum_{x \in F_q} x^u = 0 .$$

Proof: If $u = 0$,

$$\sum_{x \in F_q} x^0 = \sum_{x \in F_q} 1 = q \cdot 1 = 0 .$$

If $0 < u < q-1$, let a be a generator of the cyclic group F_q^*. Since a has order $q-1$, it follows that $a^u \neq 1$. But as x runs through F_q, then so does ax, so that

$$\sum_{x \in F_q} x^u = \sum_{x \in F_q} (ax)^u = a^u \sum_{x \in F_q} x^u .$$

The result follows immediately.

LEMMA 1B: *Suppose* $f(\underline{\underline{X}}) = f(X_1, \ldots, X_n)$ *is of total degree* $d < n(q-1)$. *Then*

$$\sum_{\underline{\underline{x}} \in F_q^n} f(\underline{\underline{x}}) = 0 .$$

Proof: By linearity, it is clear that we may restrict our attention to the case where $f(\underline{\underline{X}}) = X_1^{u_1} X_2^{u_2} \ldots X_n^{u_n}$. Then

$$\sum_{\underline{\underline{x}} \in F_q^n} f(\underline{\underline{x}}) = \prod_{i=1}^{n} \left(\sum_{x_i \in F_q} x_i^{u_i} \right) .$$

But since $u_1 + u_2 + \ldots + u_n = d < n(q-1)$, there is a u_j with $nu_j \leq d < n(q-1)$, whence with $u_j < q-1$. By Lemma 1A,

$$\sum_{x_j \in F_q} x_j^{u_j} = 0 ,$$

and the desired conclusion follows.

<u>THEOREM 1C</u>: (Warning's Theorem). <u>Let</u> F_q <u>be of characteristic</u> p . <u>Let</u> $f_1(\underline{\underline{X}}), \ldots, f_t(\underline{\underline{X}})$ <u>be polynomials in</u> $F_q[\underline{\underline{X}}]$ <u>of total degrees</u> $d_1, \ldots, d_t$, <u>respectively, and suppose that</u>

(1.1) $$d = d_1 + \ldots + d_t < n .$$

<u>Then the number</u> N <u>of common zeros of</u> $f_1, \ldots, f_t$ <u>satisfies</u>

$$N \equiv 0 \pmod{p} .$$

<u>Proof</u>: Introduce the polynomial

$$g(\underline{\underline{X}}) = (1 - f_1^{q-1}(\underline{\underline{X}})) \ldots (1 - f_t^{q-1}(\underline{\underline{X}})) .$$

Then g has total degree $d(q-1) < n(q-1)$, so that by Lemma 1B,

$$\sum_{\underline{\underline{x}} \in F_q^n} g(\underline{\underline{x}}) = 0 .$$

On the other hand, for any $\underline{\underline{x}} \in F_q^n$, we have $f_i^{q-1}(\underline{\underline{x}}) = 1$, unless $f_i(\underline{\underline{x}}) = 0$. Hence $g(\underline{\underline{x}}) = 0$, unless $\underline{\underline{x}}$ is a common zero of $f_1, \ldots, f_t$, in which case $g(\underline{\underline{x}}) = 1$. Therefore

$$0 = \sum_{\underline{\underline{x}} \in F_q^n} g(\underline{\underline{x}}) = N .$$

It follows that $N \equiv 0 \pmod{p}$.

Theorem 1C was proved by Warning in (1935). The next theorem was conjectured by E. Artin in 1934, and was proved prior to Warning's Theorem.

<u>THEOREM 1D</u>: (Chevalley (1935)). <u>Let</u> $f(\underline{\underline{X}})$ <u>be a form of degree</u> $d < n$. <u>Then</u> f <u>has a non-trivial zero in</u> F_q^n.

<u>Proof</u>: Since f has no constant term, $\underline{\underline{0}} \in F_q^n$ is a zero of f. If N is the number of zeros of f in F_q^n, then $N \geq 1$. But since $d < n$, Theorem 1C says that p divides N, so that in fact $N \geq p$. Therefore the number of non-trivial zeros of f in F_q is

$$N - 1 \geq p - 1 \geq 1 .$$

<u>Remark</u>: Theorems 1C and 1D are no longer true when $d = n$. For any positive integer n and any prime power q, let $\omega_1, \ldots, \omega_n$ be a basis of F_{q^n} over F_q. Let

$$g(\underline{\underline{X}}) = \prod_{j=0}^{n-1} (\omega_1^{q^j} X_1 + \ldots + \omega_n^{q^j} X_n) .$$

Observe that $g(\underline{\underline{X}})$ is a polynomial in n variables of total degree n. By Theorem 1E of Chapter I, the elements $\omega_i^{q^j}$ $(0 \leq j \leq n-1)$ are the conjugates of ω_i. Since $g(\underline{\underline{X}})$ is evidently invariant under the Galois group of F_{q^n} over F_q, it has coefficients in F_q. Moreover, if $\underline{\underline{x}} = (x_1, \ldots, x_n) \in F_q^n$ and $\xi = \omega_1 x_1 + \ldots + \omega_n x_n$, then $g(\underline{\underline{x}})$ is the norm $\mathfrak{N}(\xi)$ of ξ. Hence if $\underline{\underline{x}} \in F_q^n$ and $\underline{\underline{x}} \neq \underline{\underline{0}}$, then $\xi \neq 0$, whence

$$g(\underline{\underline{x}}) = \mathfrak{N}(\omega_1 x_1 + \ldots + \omega_n x_n) = \mathfrak{N}(\xi) \neq 0 .$$

Therefore $g(\underline{\underline{X}})$ has only the trivial zero. So $N = 1$ and $N \not\equiv 0 \pmod{p}$.

THEOREM 1E: (Warning's Second Theorem) (Warning (1935)). *Under the hypothesis of Theorem 1C, if* $N > 0$, *then*

$$N \geq q^{n-d} .$$

Given a subspace S of F_q^n and an element $\underline{t} \in F_q^n$, let

$$W = S + \underline{t}$$

be the set of points $\underline{s} + \underline{t}$ with $\underline{s} \in S$. Such a set W will be called a *linear manifold*. The subspace S (but not $\underline{t}$) is determined by W, and we may say that W is obtained from S by a translation. The *dimension* of W is by definition the dimension of S . Two linear manifolds of the same dimension are said to be *parallel* if they are obtained from the same subspace S .

In what follows, V will be the set of $\underline{x} \in F_q^n$ with $f_1(\underline{x}) = \ldots = f_t(\underline{x}) = 0$.

LEMMA 1F: *If* W_1 *and* W_2 *are two parallel linear manifolds, then*

$$|W_1 \cap V| \equiv |W_2 \cap V| \pmod{p} .$$

Proof: Since the case where $W_1 = W_2$ is obvious, we may assume that $W_1 \neq W_2$. Moreover, after a linear change of coordinates, we may suppose that

$$W_1 = \{(x_1, \ldots, x_n) : 0 = x_1 = x_2 = \ldots = x_{n-d}\}$$

and

$$W_2 = \{(x_1, \ldots, x_n) : 1 = x_1 , \ 0 = x_2 = \ldots = x_{n-d}\} .$$

Now write

$$r(X) = X^{q-1} - 1 = \prod_{a \in F_q^*} (X - a) ,$$

and

$$g(\underline{\underline{X}}) = (-1)^{n-d} r(X_2) \dots r(X_{n-d}) \prod_{\substack{a \neq 0,1 \\ a \in F_q}} (X_1 - a) .$$

It may be seen that $g(\underline{\underline{X}})$ is a polynomial of total degree $(n-d)(q-1)-1$, with the property that

$$g(\underline{\underline{x}}) = \begin{cases} -1 & \text{if } \underline{\underline{x}} \in W_1 , \\ 1 & \text{if } \underline{\underline{x}} \in W_2 , \\ 0 & \text{otherwise} . \end{cases}$$

Put

$$h(\underline{\underline{X}}) = (1 - f_1^{q-1}(\underline{\underline{X}})) \dots (1 - f_t^{q-1}(\underline{\underline{X}}))g(\underline{\underline{X}}) .$$

$h(\underline{\underline{X}})$ is a polynomial in n variables of total degree

$$(n-d)(q-1) - 1 + d(q-1) = n(q-1) - 1 < n(q-1) .$$

Furthermore,

$$h(\underline{\underline{x}}) = \begin{cases} -1 & \text{if } \underline{\underline{x}} \in W_1 \cap V , \\ 1 & \text{if } \underline{\underline{x}} \in W_2 \cap V , \\ 0 & \text{otherwise} . \end{cases}$$

Hence

$$\sum_{\underline{\underline{x}} \in F_z^n} h(\underline{\underline{x}}) = |W_2 \cap V| - |W_1 \cap V| .$$

But Lemma 1B is applicable to $h(\underline{\underline{X}})$, and yields

$$|W_1 \cap V| \equiv |W_2 \cap V| \pmod p .$$

<u>Proof of Theorem 1E</u>: There are two cases.

<u>Case 1</u>: There exists a linear manifold W of dimension d such that

$$|W \cap V| \not\equiv 0 \pmod p .$$

By Lemma 1F, if W' is any linear manifold of dimension d parallel to W, then

(1.2) $$|W' \cap V| \not\equiv 0 \pmod{p} .$$

There are exactly q^{n-d} parallel linear manifolds (including W itself), and they form a partition of F_q^n . Since by (1.2) each contains at least one point of V, we have

$$N = |V| \geq q^{n-d} .$$

<u>Case 2</u>: For all linear manifolds W of dimension d,

$$|W \cap V| \equiv 0 \pmod{p} .$$

Since by hypothesis, V contains at least one point, there exists an integer m, $1 \leq m \leq d$, with two properties:

(i) For every linear manifold M of dimension m,

$$|M \cap V| \equiv 0 \pmod{p} .$$

(ii) There is a linear manifold L of dimension $m-1$ such that

$$|L \cap V| \not\equiv 0 \pmod{p} .$$

Fix one such linear manifold L.

Given a set A and a subset B, write $A \sim B$ for the complement of B in A. Consider the linear manifolds M of dimension m containing L; of these there are exactly

$$\frac{q^{n-m+1} - 1}{q - 1} = q^{n-m} + \ldots + q + 1 .$$

We have $|M \cap V| \equiv 0 \pmod{p}$ but $|L \cap V| \not\equiv 0 \pmod{p}$, whence $|(M \sim L) \cap V| \not\equiv 0 \pmod{p}$ and

$$|(M \sim L) \cap V| \geq 1 .$$

But the sets $M \sim L$ form a partition of $F_q^n \sim L$; thus

$$N = |V| > q^{n-m} + \ldots + q + 1 > q^{n-d} .$$

THEOREM $1G^*$: (J. Ax (1964)) *Make the same hypotheses as in Theorem 1C. Let b be an integer, $b < n/d$. Then*

$$N \equiv 0 \pmod{q^b} .$$

This is a great improvement over Theorem 1C. The proof of this theorem will not be included in these lectures. See Ax's original paper or Joly (1973), Chapter 7.

§2. Quadratic forms.

Let K be a field whose characteristic is not 2. A *quadratic form* f over K is a polynomial over K of the type

$$f(\underline{\underline{X}}) = f(X_1,\dots,X_n) = \sum_{1\le i,k\le n} a_{ik} X_i X_k ,$$

where $a_{ik} = a_{ki}$. The *determinant of* f, abbreviated det f, is the determinant of the $(n \times n)$-matrix of coefficients of f: det f = det (a_{ik}). We say that $f(\underline{\underline{X}})$ is *nondegenerate* if det f $\neq 0$.
Let M^t denote the transpose of a matrix M. If we take

$$A = \begin{pmatrix} a_{11} & a_{12} & \cdots & a_{1n} \\ a_{21} & & & \vdots \\ \vdots & & & \vdots \\ a_{n1} & \cdots & \cdots & a_{nn} \end{pmatrix} , \quad \underline{\underline{X}} = \begin{pmatrix} X_1 \\ X_2 \\ \vdots \\ X_n \end{pmatrix} , \quad \underline{\underline{X}}^t = (X_1, X_2 \dots X_n) ,$$

then $A = A^t$ and $f(\underline{\underline{X}}) = \underline{\underline{X}}^t A \underline{\underline{X}}$.

Now let $f(\underline{\underline{X}})$ and $g(\underline{\underline{X}})$ be two quadratic forms over K. We say that $f(\underline{\underline{X}})$ is *equivalent* to $g(\underline{\underline{X}})$, written $f(\underline{\underline{X}}) \sim g(\underline{\underline{X}})$, if there is a non-singular matrix T such that $g(\underline{\underline{X}}) = f(T\underline{\underline{X}})$.

It is clear that "$\sim$" is an equivalence relation. If f has the matrix A, and if $g(\underline{X}) = f(T\underline{X})$, then g has the matrix T^tAT and

$$\det g = \det f \cdot (\det T)^2 .$$

If $f(\underline{X}) \sim g(\underline{X})$ and $f(\underline{X})$ is nondegenerate, then $g(\underline{X})$ is also nondegenerate and $\det f/\det g \in (K^*)^2$; that is, $\det f/\det g$ is a non-zero square in K.

Suppose $a \in K$, $a \neq 0$. We say that a quadratic form $f(\underline{X})$ <u>represents</u> a if there are $x_1,\ldots,x_n$ in K so that $f(x_1,\ldots,x_n) = a$. We say $f(\underline{X})$ <u>represents zero</u> if there are $x_1,\ldots,x_n$ in K, with $(x_1,\ldots,x_n) \neq (0,\ldots,0)$, such that $f(x_1,\ldots,x_n) = 0$. Clearly, equivalent forms represent the same elements of K.

<u>LEMMA 2A</u>: <u>Suppose that a quadratic form $f(\underline{X})$ represents a non-zero element $a \in K$. Then for some quadratic form g in $n-1$ variables,</u>

$$f(X_1,\ldots,X_n) \sim aX_1^2 + g(X_2,\ldots,X_n) .$$

<u>Proof</u>: Let A be the matrix of coefficients of $f(\underline{X})$. By hypothesis, there exists an $\underline{x} \in K^n$ with $f(\underline{x}) = \underline{x}^t A \underline{x} = a$. Since $\underline{x} \neq \underline{0}$, it is clearly possible to select a non-singular matrix

$$C = \begin{pmatrix} x_1 & c_{12} & \cdots & c_{1n} \\ \vdots & & & \vdots \\ x_n & c_{n2} & \cdots & c_{nn} \end{pmatrix}$$

with entries in K. Now $f(C\underline{X}) = \underline{X}^t C^t A C \underline{X}$, and it is easy to see that the entry in the upper left corner of C^tAC is $\underline{x}^t A \underline{x} = a$.

Therefore for certain $b_2,\ldots,b_n$,

$$f(X_1,\ldots,X_n) \sim aX_1^2 + 2b_2X_1X_2 + \ldots + 2b_nX_1X_n + h(X_2,\ldots,X_n)$$

$$= a\left|X_1 + (b_2/a)X_2 + \ldots + (b_n/a)X_n\right|^2 + g(X_2,\ldots,X_n) .$$

After making the non-singular transformation $X' = X_1 + (b_2/a)X_2 + \ldots + (b_n/a)X_n$, $X_2' = X_2$, ..., $X_n' = X_n$, we see that

$$f(X_1,\ldots,X_n) \sim aX_1^2 + g(X_2,\ldots,X_n) .$$

A quadratic form $f(\underline{\underline{X}})$ is called <u>diagonal</u> if $f(\underline{\underline{X}}) = a_1X_1^2 + \ldots a_nX_n^2$.

<u>LEMMA 2B</u>: <u>Every quadratic form is equivalent to a diagonal form.</u>

<u>Proof</u>: The proof is by induction on n. If $n = 1$, then $f(\underline{\underline{X}}) = a_{11}X_1^2$ is always in diagonal form. Suppose the lemma holds for forms in $n - 1$ variables. Let $f(\underline{\underline{X}}) = f(X_1,\ldots,X_n)$ be a form in n variables. The lemma is true if $f(\underline{\underline{X}}) = 0$. Otherwise either some $a_{ii} \neq 0$, in which case f represents $a_{ii} \neq 0$. Or all a_{ii} are zero, but some $a_{ij} = a_{ji} \neq 0$. Then f represents $2a_{ij}$, since $f(0,\ldots,\underset{i}{1},\ldots,\underset{j}{1},\ldots,0) = 2a_{ij}$. Hence f represents some non-zero element a, and

$$f \sim aX_1^2 + g(X_2,\ldots,X_n)$$

by Lemma 2A. By induction, $g \sim a_2X_2^2 + \ldots + a_nX_n^2$, and

$$f \sim aX_1^2 + a_2X_2^2 + \ldots + a_nX_n^2 .$$

LEMMA 2C: If a nondegenerate quadratic form represents zero, then it represents every element of the field K.

Proof: Let $f(\underline{X})$ be a nondegenerate quadratic form over K which represents zero. By using equivalence, we may suppose that $f(\underline{X})$ is diagonal:

$$f(\underline{X}) = f(X_1,\ldots,X_n) = a_1X_1^2 + \ldots + a_nX_n^2 .$$

Since $f(\underline{X})$ is nondegenerate, $a_1 \neq 0, \ldots, a_n \neq 0$. Since $f(\underline{X})$ represents zero, there exist $n \geq 2$ elements $x_1,\ldots,x_n$ in K, not all zero, with

$$f(\underline{x}) = f(x_1,\ldots,x_n) = a_1x_1^2 + \ldots + a_nx_n^2 = 0 .$$

Without loss of generality, we may assume $x_1 \neq 0$. Put $y_1 = x_1(1+t)$, $y_2 = x_2(1-t), \ldots, y_n = x_n(1-t)$, with $t \in K$ to be determined. Then

$$f(y_1,\ldots,y_n) = 2t(a_1x_1^2 - a_2x_2^2 - \ldots - a_nx_n^2)$$

$$= 4ta_1x_1^2 .$$

Now if $a \in K^*$ and if we set $t = a/(4a_1x_1^2)$, we obtain $f(y_1,\ldots,y_n) = a$ Thus f represents a.

We now return to our general theme by focusing attention on quadratic forms over a finite field. Since it was necessary that we require char $K \neq 2$ in this section, we consider finite fields F_q with q odd. Suppose $d \in F_q^*$. We introduce the notation:

$$\left(\frac{d}{q}\right) = \begin{cases} 1 & \text{if} \quad d \in (F_q^*)^2 , \\ -1 & \text{if} \quad 1 \notin (F_q^*)^2 . \end{cases}$$

Suppose $f_1(\underline{X})$ and $f_2(\underline{X})$ are equivalent nondegenerate quadratic forms over F_q with respective determinants d_1 and d_2. Then $d_1/d_2 \in (F_q^*)^2$, whence $\left(\frac{d_1}{q}\right) = \left(\frac{d_2}{q}\right)$. That is, the symbol $\left(\frac{d}{q}\right)$ is invariant under equivalence.

LEMMA 2D: *Let* $f(X_1,\ldots,X_n)$, $n \geq 3$, *be a nondegenerate quadratic form over* F_q, *where* q *is odd. Then*

$$f(X_1,\ldots,X_n) \sim X_1X_2 + h(X_3,\ldots,X_n) .$$

Proof: By Chevalley's Theorem (Theorem 1D), $f(\underline{X})$ has a non-trivial zero in F_q; i.e., $f(\underline{X})$ represents zero. By Lemma 2C, $f(\underline{X})$ represents $1 \in F_q$. By Lemma 2A, $f(\underline{X}) \sim X_1^2 + g(X_2,\ldots,X_n)$ for some form g. Hence $X_1^2 + g(X_2,\ldots,X_n)$ represents zero, so there exist $x_1,\ldots,x_n \in F_q$, not all zero, with

$$x_1^2 + g(x_2,\ldots,x_n) = 0 .$$

If $x_1 \neq 0$, then g represents $-x_1^2$, hence g represents -1. If $x_1 = 0$, then g represents zero, and therefore, by Lemma 2C, g again represents -1. By Lemma 2A,

$$g(X_2,\ldots,X_n) \sim -X_2^2 + h(X_3,\ldots,X_n) ,$$

whence

$$\begin{aligned} f(X_1,\ldots,X_n) &\sim X_1^2 - X_2^2 + h(X_3,\ldots,X_n) \\ &\sim X_1X_2 + h(X_3,\ldots,X_n) . \end{aligned}$$

Now let N_n be the number of zeros in F_q^n of $f(X_1,\dots,X_n)$, and let N_{n-2} be the number of zeros in F_q^{n-2} of $h(X_3,\dots,X_n)$. In order to find the relation between N_n and N_{n-2}, we observe that $f(X_1,\dots,X_n)$ and $X_1X_2 + h(X_3,\dots,X_n)$ must have the same number of zeros, since they are equivalent.

We first count solutions of

$$x_1x_2 + h(x_3,\dots,x_n) = 0$$

with $h(x_3,\dots,x_n) = 0$, hence with $x_1x_2 = 0$. The number of possibilities for $x_3,\dots,x_n$ is N_{n-2}, the number of possibilities for x_1,x_2 is $2q - 1$, so that altogether we obtain

$$(2q - 1)\, N_{n-2} .$$

We next count solutions with $h(x_3,\dots,x_n) \neq 0$. The number of possibilities for $x_3,\dots,x_n$ is $q^{n-2} - N_{n-2}$, and for given $x_3,\dots x_n$, the number of possibilities for x_1,x_2 is $q - 1$, so that we get

$$(q - 1)(q^{n-2} - N_{n-2})$$

such solutions. Adding these two numbers, we obtain

$$N_n = q^{n-1} - q^{n-2} + qN_{n-2} . \tag{2.1}$$

<u>THEOREM 2E</u>: <u>Let</u> $f(\underline{\underline{X}}) = f(X_1,\dots,X_n)$ <u>be a nondegenerate quadratic form of determinant</u> d <u>over</u> F_q, q <u>odd. Then the number</u> N <u>of zeros of</u> $f(\underline{\underline{X}})$ <u>in</u> F_q <u>is given by</u>

$$(2.2) \qquad N = \begin{cases} q^{n-1}, & \text{\underline{if} } n \text{ \underline{is odd},} \\ q^{n-1} + (q-1)q^{(n-2)/2}\left(\dfrac{(-1)^{n/2}d}{q}\right), & \text{\underline{if} } n \text{ \underline{is even}.} \end{cases}$$

Proof: Suppose n is odd. If $n = 1$, $f(X) = aX^2$, and $N = 1$. If $n \geq 3$, we may suppose that $f = X_1X_2 + h(X_3,\ldots,X_n)$. If the theorem holds for $n - 2$, then $N_{n-2} = q^{n-3}$ and by (2.1),

$$\begin{aligned} N &= q^{n-1} - q^{n-2} + qN_{n-2} \\ &= q^{n-1} - q^{n-2} + q \cdot q^{n-3} \\ &= q^{n-1} . \end{aligned}$$

Now suppose n is even. If $n = 2$, $f(X_1,X_2)$ is equivalent to a nondegenerate diagonal form

$$a_1X_1^2 + a_2X_2^2 = a_1(X_1^2 + (a_2/a_1)X_2^2) ,$$

and $\left(\frac{-d}{q}\right) = \left(\frac{-a_1a_2}{q}\right)$. If $\left(\frac{-d}{q}\right) = -1$, then $\left(\frac{-a_1a_2}{q}\right) = -1$, whence $\left(\frac{-(a_2/a_1)}{q}\right) = -1$. If (x_1,x_2) were a non-trivial zero of $f(X_1,X_2)$, then $x_1^2 = -(a_2/a_1)x_2^2$, which is impossible. Therefore $f(X_1,X_2)$ has only the trivial zero; i.e. $N = 1$, which agrees with (2.2). If $\left(\frac{-d}{q}\right) = +1$, then in a similar way $\left(\frac{-(a_2/a_1)}{q}\right) = +1$, and we see that $x_1^2 = -(a_2/a_1)x_2^2$ has $2(q-1)$ non-trivial solutions (x_1,x_2). Therefore $N = 1 + 2(q-1) = 2q - 1$, again agreeing with (2.2).

If $n \geq 4$, we may suppose that $f = X_1X_2 + h(X_3,\ldots,X_n)$. Observe that the determinant of h is minus that of f. Now suppose the theorem holds for $n - 2$. Then

$$N_n = q^{n-1} - q^{n-2} + qN_{n-2}$$

$$= q^{n-1} - q^{n-2} + q\left(q^{n-3} + (q-1)q^{(n-4)/2}\left(\frac{(-1)^{(n-2)/2}(-d)}{q}\right)\right)$$

$$= q^{n-1} + (q-1)q^{(n-2)/2}\left(\frac{(-1)^{n/2}d}{q}\right) .$$

§3. Elementary upper bounds. Projective zeros.

LEMMA 3A: *Let* $f(X_1,\ldots,X_n)$ *be a non-zero polynomial over* F_q *of total degree* d. *Then the number* N *of zeros of* $f(X_1,\ldots,X_n)$ *in* F_q^n *satisfies*

$$N \le dq^{n-1} .$$

If $f(X_1,\ldots,X_n)$ *is homogeneous, then the number of its non-trivial zeros is at most* $d(q^{n-1} - 1)$.

Proof: If $d = 0$, f is a non-zero constant and has no zeros. If $d = 1$, then

$$f(X_1,\ldots,X_n) = a_1X_1 + \ldots + a_nX_n + c ,$$

and $N = q^{n-1}$. If f is homogeneous of degree $d = 1$, then $c = 0$ and the number of non-trivial zeros of f is $q^{n-1} - 1$. If $n = 1$, then clearly $N \le d$. If $n = 1$ and f is homogeneous, then f can have no non-trivial zeros.

We have shown that the lemma holds if $d \le 1$ or if $n = 1$. We proceed by "double induction". Suppose $n > 1$, $d > 1$, and the lemma is true for polynomials in at most n variables of degree less than d, and the lemma is true for polynomials in less than n variables of degree at most d. We must prove the lemma for a polynomial $f(X_1,\dots,X_n)$ in n variables of degree d. There are two cases.

<u>Case 1</u>: $f(X_1,\dots,X_n)$ is not divisible by $X_1 - x$ for any $x \in F_q$. Then for any $x \in F_q$, $f(x,X_2,\dots,X_n)$ is a non-zero polynomial of degree at most d in $n-1$ variables. By the inductive hypothesis, the number of zeros $(x_2,\dots,x_n) \in F_q^{n-1}$ of $f(x,X_2,\dots,X_n)$ is at most dq^{n-2}. But we have q choices for $x \in F_q$, so that $N \le qdq^{n-2} = dq^{n-1}$.

By the same reasoning, the number of zeros of $f(X_1,X_2,\dots,X_n)$ with $x_1 \neq 0$ is at most $(q-1)dq^{n-2}$. If $f(X_1,\dots,X_n)$ is homogeneous, then so is $f(0,X_2,\dots,X_n)$, and the number of non-trivial zeros of $f(0,X_2,\dots,X_n)$ is at most $d(q^{n-2}-1)$ by induction. Therefore the total number of non-trivial zeros of $f(X_1,\dots,X_n)$ is

$$\le d(q-1)q^{n-2} + d(q^{n-2}-1)$$

$$= d(q^{n-1}-1) .$$

<u>Case 2</u>: $f(X_1,\dots,X_n)$ is divisible by $X_1 - x$ for some $x \in F_q$. Then $f(\underline{\underline{X}}) = (X_1 - x)g(\underline{\underline{X}})$, where g is a non-zero polynomial in at most n variables of degree at most $d-1$. We immediately see that

$$N \le q^{n-1} + (d-1)q^{n-1} = dq^{n-1}.$$

If f is homogeneous, then necessarily $x = 0$ and $f(\underline{\underline{X}}) = X_1 g(\underline{\underline{X}})$. The number of non-trivial zeros of f is

$$\leq (q^{n-1} - 1) + (d - 1)(q^{n-1} - 1)$$

$$= d(q^{n-1} - 1) .$$

Remark: If $f(\underline{X}) = (X_1 - c_1)(X_1 - c_2) \ldots (X_1 - c_d)$ where $c_1, c_2, \ldots, c_d$ are distinct elements of F_q, then $N = dq^{n-1}$. However, for homogeneous polynomials, our estimate is in general not best possible

K^n, where K is a field, is called n-dimensional space over K, or more precisely, n-dimensional affine space over K. On the other hand, n-dimensional projective space over K by definition consists of non-zero $(n + 1)$ - tuples $(x_0, x_1, \ldots, x_n)$ with components in K, and with proportional $(n + 1)$ - tuples considered equal. A point in projective space is called "finite" if it is represented by $(x_0, x_1, \ldots, x_n)$ with $x_0 \neq 0$. Every finite point of projective space may be uniquely represented by some $(1, y_1, \ldots, y_n)$. Hence there is a 1-1 correspondence between finite points of projective space and points of affine space. Points of projective space represented by $(0, x_1, \ldots, x_n)$ are called "infinite points", or "points at infinity".

Now suppose $f(\underline{X})$ is a polynomial of degree $d > 0$, say

$$f(\underline{X}) = f(X_1, \ldots, X_n) = \sum_{i_1 + \ldots + i_n \leq d} a_{i_1, i_2, \ldots, i_n} X_1^{i_1} X_2^{i_2} \ldots X_n^{i_n} .$$

Associate with $f(\underline{X})$ the form

$$f^*(X_0, X_1, \ldots, X_n) = \sum_{i_0 + i_1 + \ldots + i_n = d} a_{i_1, i_2, \ldots, i_n} X_0^{i_0} X_1^{i_1} \ldots X_n^{i_n} .$$

We may say that the equation $f(\underline{\underline{X}}) = 0$ defines a "hypersurface in n-space". The zeros of $f(\underline{\underline{X}})$ are the "points" of this hypersurface. The equation $f^*(x_0,x_1,\ldots,x_n) = 0$ defines a "hypersurface in n-dimensional projective space". In this case, we consider only non-trivial zeros $(x_0,x_1,\ldots,x_n) \neq (0,0,\ldots,0)$, and two zeros are considered identical if their coordinates are proportional. These are called "points on the projective hypersurface", or "projective zeros".

Suppose $(x_0,x_1,\ldots,x_n)$ represents a zero of f^*. There are two possibilities:

(a) $x_0 \neq 0$. The zero may then be represented uniquely by an $(n+1)$ - tuple $(1,y_1,\ldots,y_n)$. Since $f^*(1,y_1,\ldots,y_n) = 0$, we have $f(y_1,\ldots,y_n) = 0$. Conversely, if $(y_1,\ldots,y_n)$ is a zero of f, then $(1,y_1,\ldots,y_n)$ is a zero of f^*. These points of the projective hypersurface are called "finite". There is thus a 1-1 correspondence between finite points on the projective hypersurface $f^* = 0$ and points on the affine hypersurface $f = 0$.

(b) $x_0 = 0$. These points are called "points at infinity" of the hypersurface.

<u>Example</u>: Let $f(X_1,X_2) = X_1^2 - X_2^2 - 1$. The equation $f(x_1,x_2) = 0$ defines a hyperbola. This hyperbola has the two asymptotes $x_2 = x_1$ and $x_2 = -x_1$. In this example, $f^*(X_0,X_1,X_2) = X_1^2 - X_2^2 - X_0^2$. The points at infinity are the zeros of f^* with $x_0 = 0$. There are, if char $K \neq 2$, two points at infinity, represented by $(0,1,1)$ and $(0,1,-1)$. They may be interpreted as "points infinitely far out on the two asymptotes".

Whether or not there exist points at infinity may depend on the underlying field.

Example: Let $f(X_1,X_2) = X_1^2 + X_2^2 - 1$. The equation $f(x_1,x_2) = 0$ defines a circle of radius 1 . Since here $f^*(X_0,X_1,X_2) = X_1^2 + X_2^2 - X_0^2$, the points at infinity are those elements $(0,x_1,x_2)$ satisfying $x_1^2 + x_2^2 = 0$. If the field under consideration is the field $\mathbb{R}$ of reals, there is no point at infinity. If our field is the field $\mathbb{C}$ of complex numbers, there are two points at infinity represented by $(0,1,i)$ and $(0,1,-i)$.

LEMMA 3B: Let $f(\underline{X})$ be a polynomial of degree d with coefficients in F_q . Let N be the number of zeros of f in F_q^n . Let N^* be the number of projective zeros as defined above. Then

$$N \le N^* \le N + d(q^{n-2} + q^{n-3} + \dots + q + 1) .$$

Proof: Since N^* is the sum of N and the number of points at infinity, we have $N \le N^*$, and we simply have to estimate the number of points at infinity. The number of non-trivial zeros of $f^*(0,X_1,\dots,X_n)$ is at most $d(q^{n-1} - 1)$ by Lemma 3A. But two such zeros are considered identical when they are proportional, so that the number of points at infinity is at most

$$d(q^{n-1} - 1)/(q - 1) = d(q^{n-2} + q^{n-3} + \dots + q + 1) .$$

The lemma follows.

Exercise. Show that $f^*(X_0,X_1,\dots,X_n)$ is irreducible precisely if f is.

LEMMA 3C: Suppose $n \geq 2$. Let $u_1(X_1,\ldots,X_n)$ and $u_2(X_1,\ldots,X_n)$ be polynomials over F_q of respective total degrees e_1 and e_2, without common factor of positive degree. Then the number of their common zeros in F_q^n is at most

$$q^{n-2} e_1 e_2 \min\{e_1, e_2\} .$$

Remark: The estimate of Lemma 3C is not best possible.

Proof of Lemma 3C: Without loss of generality, suppose $e_1 \leq e_2$, so that $e_1 = \min\{e_1, e_2\}$. If $e_1 = 0$, then u_1, is constant. If $u_1(\underline{X}) = c \neq 0$, there are no common zeros, and the lemma holds. If $u_1(\underline{X}) = 0$, then $u_2(\underline{X})$ is a non-zero constant (otherwise $u_1(\underline{X})$ and $u_2(\underline{X})$ would have a common factor), and again there are no common zeros. If $e_1 = 1$, then $u_1(\underline{X})$ is linear. After an appropriate linear transformation, we may suppose $u_1(\underline{X}) = X_1$. If $\underline{x} = (x_1,\ldots,x_n)$ is a common zero; i.e., $u_1(\underline{x}) = u_2(\underline{x}) = 0$, then $x_1 = 0$ and $u_2(0,x_2,\ldots,x_n) = 0$. But $u_2(0,X_2,\ldots,X_n) \neq 0$, so that by Lemma 3A the number of common zeros is at most $e_2 q^{n-2}$, agreeing with the estimate of the lemma when $e_1 = 1$.

Now suppose $e_1 \geq 2$. Every common zero of $u_1(\underline{X})$ and $u_2(\underline{X})$ is a zero of $u_1(\underline{X})$, so the number of common zeros is certainly

$$\leq e_1 q^{n-1}$$

$$\leq q^{n-2} e_1 e_2 \min\{e_1, e_2\}$$

if $q \leq e_1 e_2$. We may then suppose that $q > e_1 e_2 \geq e_1 + e_2$. Let

$$v_j(X_1,\ldots,X_n) = u_j(X_1, X_2 + c_2X_1,\ldots,X_n + c_nX_1)$$
$$= p_j(c_2,\ldots,c_n)X_1^{e_j} + \ldots \quad .$$

We wish to choose $c_2,\ldots, c_n \in F_q$ so that the coefficient of $X_1^{e_1}$ in $v_1(\underline{X})$ and of $X_1^{e_2}$ in $v_2(\underline{X})$ are not zero. Now p_j is a polynomial of degree at most e_j, and is not identically zero. By Lemma 3A, the total number of zeros of p_j in F_q^{n-1} is at most e_jq^{n-2}. Therefore the total number of zeros of both p_1 and p_2 is

$$\leq (e_1 + e_2)q^{n-2} < q^{n-1} \quad .$$

It is therefore possible to choose $(c_2,\ldots,c_n) \in F_q^{n-1}$ with $p_1(c_2,\ldots,c_n) \neq 0$ and $p_2(c_2,\ldots,c_n) \neq 0$. Hence after a non-singular linear transformation, and after division by $p_1(c_2,\ldots,c_n)$ and $p_2(c_2,\ldots,c_n)$, respectively, we may assume without loss of generality that

$$u_1(\underline{X}) = X_1^{e_1} + X_1^{e_1-1}g_1(X_2,\ldots,X_n) + \ldots + g_{e_1}(X_2,\ldots,X_n) ,$$

$$u_2(\underline{X}) = X_1^{e_2} + X_1^{e_2-1}h_1(X_2,\ldots,X_n) + \ldots + h_{e_2}(X_2,\ldots,X_n) .$$

Considering $u_1(\underline{\underline{X}})$ and $u_2(\underline{\underline{X}})$ as polynomials in X_1, their resultant is a polynomial $R(X_2,\ldots,X_n)$. It is not hard to see that the total degree of R is at most e_1e_2. But by the basic property of resultants, for any common zero $(x_1,x_2,\ldots,x_n)$ of $u_1(\underline{\underline{X}})$ and $u_2(\underline{\underline{X}})$, $R(x_2,\ldots,x_n) = 0$. The number of such $(n-1)$-tuples $(x_2,\ldots,x_n)$ is at most $e_1e_2q^{n-2}$ by Lemma 3A, and for such $x_2,\ldots,x_n$, the number of possibilities for x_1 is clearly not more than e_1. So the total number of common zeros of $u_1(\underline{\underline{X}})$ and $u_2(\underline{\underline{X}})$ is

$$\le q^{n-2}e_1e_2e_1$$
$$= q^{n-2}e_1e_2\min\{e_1,e_2\} .$$

<u>LEMMA 3D</u>: <u>Let $u_1(\underline{\underline{X}}),\ldots,u_t(\underline{\underline{X}})$ be polynomials in n variables over F_q, each of total degree at most e, and without common factor. Then the number of their common zeros is at most</u>

$$q^{n-2}e^3 .$$

<u>Proof</u>. The proof is by induction on t. The case $t = 2$ is Lemma 3C. Suppose $t \ge 3$, and the lemma holds for $t - 1$. Let[†)]

$$v(\underline{\underline{X}}) = \text{g.c.d.}(u_1(\underline{\underline{X}}),\ldots,u_{t-1}(\underline{\underline{X}})) ,$$

and $d = \deg v(\underline{\underline{X}})$. Then

[†)] We are implicitly using the fact that polynomials in n variables over a field form a Unique Factorization Domain.

$$u_i(\underline{X}) = v(\underline{X}) w_i(\underline{X}) \qquad (i = 1,2,\ldots,t-1) \quad ,$$

where $\deg w_i(\underline{X}) \leq e - d$, and where $w_1,\ldots,w_{t-1}$ have no common factor.

Any common zero of $u_1, u_2, \ldots, u_t$ is either a common zero of v and u_t , or of $w_1,\ldots,w_{t-1}$. The number of common zeros of v and u_t is at most $d^2 e q^{n-2}$ by Lemma 3C, since $\text{g.c.d.}(v,u_t) = 1$ The number of common zeros of $w_1,\ldots,w_{t-1}$ is at most $(e - d)^3 q^{n-2}$ by the induction hypothesis. Therefore the total number of common zeros is at most

$$d^2 e q^{n-2} + (e - d)^3 q^{n-2} \leq e^3 q^{n-2} \quad .$$

Lemma 3C is not best possible. We can do better if there are only two variables:

<u>LEMMA 3E.</u> <u>Suppose</u> $u_1(X,Y)$ <u>and</u> $u_2(X,Y)$ <u>are polynomials with coefficients in a field</u> K , <u>and with no common factor of positive degree. Let</u> e_1 <u>be the total degree of</u> $u_1(X\ Y)$, e_2 <u>the total degree of</u> $u_2(X,Y)$. <u>Then the number of common zeros of</u> u_1 <u>and</u> u_2 , <u>i.e.,</u> $(x,y) \in K^2$ <u>with</u> $u_1(x,y) = u_2(x,y) = 0$, <u>is at most</u> $e_1 e_2$.

<u>Proof:</u> If u_1, u_2 have no common factor in K , then they have no common factor in $\bar{K}$. Therefore we may assume that $|K| = \infty$.

Set

$$v_j(X,Y) = u_j(X + cY, Y) \qquad (j = 1,2) \quad ,$$

where $c \in K$ is to be determined. In $v_j(X,Y)$, the term Y^{e_j} has a coefficient which is a non-zero polynomial $p_j(c)$ in c, with $\deg p_j \le e_j$. Suppose $(x_1,y_1),\ldots,(x_\nu,y_\nu)$ are distinct common zeros of u_1,u_2. Then $(x_1 - cy_1,y_1),\ldots,(x_\nu - cy_\nu,y_\nu)$ are common zeros of v_1,v_2. Since K is infinite, we may choose $c \in K$ such that

(i) if $i \ne j$, then $x_i - cy_i \ne x_j - cy_j$,

(ii) $p_1(c) \ne 0$ and $p_2(c) \ne 0$.

Then v_1 and v_2 have common zeros $(z_1,y_1),\ldots,(z_\nu,y_\nu)$, where $z_j = x_j - cy_j$ and where $z_1,\ldots,z_\nu$ are distinct. After dividing by suitable constants (namely $p_1(c)$ and $p_2(c)$), we may suppose that

$$v_1(X,Y) = Y^{e_1} + h_1(X)Y^{e_1-1} + \ldots + h_{e_1}(X),$$

$$v_2(X,Y) = Y^{e_2} + k_1(X)Y^{e_2-1} + \ldots + k_{e_2}(X).$$

Let $R(X)$ be the resultant of v_1 and v_2 when considered as polynomials in Y. $R(X)$ is a polynomial in X of degree at most e_1e_2. Since $R(z_1) = \ldots = R(z_\nu) = 0$, we obtain $\nu \leqq e_1e_2$.

Remark: Our Lemma 3E is related to a special case of Bezout's Theorem. See Van der Waerden (1955), Ch. 11.

LEMMA 3F. *Suppose $u_1(X,Y),\ldots,u_t(X,Y)$ are $t \ge 2$ polynomials over a field K without a common factor of positive degree. Suppose each polynomial has total degree at most e. Then the number of* common zeros is at most e^2 .

Proof. Exercise.

§4. The average number of zeros of a polynomial.

Let d be a positive integer. Let Ω_d be the set of all polynomials in n variables over F_q of total degree at most d. Let ω_d be the set of all n-tuples of non-negative integers $(i_1, i_2, \ldots, i_n)$ with $i_1 + i_2 + \ldots + i_n \leq d$. It is easily seen that

$$|\omega_d| = \binom{n+d}{d} .$$

If $f(\underline{\underline{X}}) \in \Omega_d$, we may write

$$f(\underline{\underline{X}}) = \sum_{(i_1,\ldots,i_n)\in\omega_d} a_{i_1,i_2,\ldots,i_n} X_1^{i_1} X_2^{i_2} \ldots X_n^{i_n} .$$

Therefore $|\Omega_d| = q^{|\omega_d|}$. For any polynomial $f(\underline{\underline{X}}) \in \Omega_d$, let $N(f)$ denote the number of zeros of $f(\underline{\underline{X}})$ in F_q^n .

THEOREM 4A:

$$\frac{1}{|\Omega_d|} \sum_{f\in\Omega_d} N(f) = q^{n-1} .$$

Proof:

$$\sum_{f \in \Omega_d} N(f) = \sum_{f \in \Omega_d} \sum_{\substack{\underline{\underline{x}} \in F_q^n \\ f(\underline{\underline{x}})=0}} 1$$

$$= \sum_{\underline{\underline{x}} \in F_q^n} \sum_{\substack{f \in \Omega_d \\ f(\underline{\underline{x}})=0}} 1$$

$$= \sum_{\underline{\underline{x}} \in F_q^n} q^{|\omega_d| - 1}$$

$$= q^n q^{|\omega_d| - 1}$$

$$= |\Omega_d| \, q^{n-1} \quad .$$

THEOREM 4B:

$$\frac{1}{|\Omega_d|} \sum_{f \in \Omega_d} (N(f) - q^{n-1})^2 = q^{n-1} - q^{n-2} \ .$$

Proof: First,

$$\sum_{f \in \Omega_d} N^2(f) = \sum_{f \in \Omega_d} \sum_{\substack{\underline{\underline{x}} \\ f(\underline{\underline{x}})=0}} \sum_{\substack{\underline{\underline{y}} \\ f(\underline{\underline{y}})=0}} 1$$

$$= \sum_{\underline{\underline{x}}, \underline{\underline{y}}} \sum_{\substack{f \in \Omega_d \\ f(\underline{\underline{x}}) = f(\underline{\underline{y}}) = 0}} 1 \ .$$

The conditions $f(\underline{\underline{x}}) = f(\underline{\underline{y}}) = 0$ are two linear equations for the coefficients of f. These two equations have rank 2 and hence have $q^{|\omega_d| - 2}$ solutions if $\underline{\underline{x}} \neq \underline{\underline{y}}$, and they have rank 1 and $q^{|\omega_d| - 1}$ solutions if $\underline{\underline{x}} = \underline{\underline{y}}$. Hence

$$\sum_{f \in \Omega_d} N^2(f) = \sum_{\underline{\underline{x}} \neq \underline{\underline{y}}} |\Omega_d| q^{-2} + \sum_{\underline{\underline{x}}} |\Omega_d| q^{-1}$$

$$= q^n(q^n - 1)|\Omega_d| q^{-2} + q^n |\Omega_d| q^{-1}$$

$$= |\Omega_d| (q^{2n-2} - q^{n-2} + q^{n-1}) .$$

Using this formula and Theorem 4A ,

$$\sum_{f \in \Omega_d} (N(f) - q^{n-1})^2 = \sum_{f \in \Omega_d} N^2(f) - 2q^{n-1} \sum_{f \in \Omega_d} N(f) + q^{2n-2} \sum_{f \in \Omega_d} 1$$

$$= |\Omega_d| (q^{2n-2} - q^{n-2} + q^{n-1}) - 2q^{n-1} |\Omega_d| q^{n-1}$$
$$+ q^{2n-2} |\Omega_d|$$
$$= |\Omega_d| (q^{n-1} - q^{n-2}) .$$

Theorem 4B tells us that the "average value" of $(N(f) - q^{n-1})^2$ is $q^{n-1} - q^{n-2} = O(q^{n-1})$. One might expect that it be often the case that

$$N(f) - q^{n-1} = 0(q^{(n-1)/2}) \quad .$$

In fact, we have shown (Theorem 1A, Chapter III) that when $n = 2$ and f is absolutely irreducible, then

$$N(f) - q = 0(q^{1/2}) \quad .$$

P. Deligne (to appear) proved that

$$N(f) - q^{n-1} = 0(q^{(n-1)/2})$$

if f is "non-singular." In fact, Deligne proved more. He proved Weil's (1949) famous conjecture on the zeta function of varieties. In the present lectures we shall not be able to prove Deligne's deep result.

§5. Additive Equations: A Chebychev Argument.

Consider a polynomial equation of the type

$$(5.1) \qquad a_1x_1^{d_1} + a_2x_2^{d_2} + \dots + a_nx_n^{d_n} = 1 \quad ,$$

where $a_i \in F_q^*$ and $d_i > 0$ $(i = 1,2,\dots,n)$.

THEOREM 5A. The number N of solutions of (5.1) in F_q^n satisfies

$$|N - q^{n-1}| \le d_1d_2 \dots d_nq^{(n-1)/2}(1 - \frac{1}{q})^{-n/2}$$

Remark: The error term here and in Theorem 5C below can be slightly improved by using exponential sums, as will be explained in §6.

Proof of the theorem: By the argument used in §2, Chapter I, the number of solutions is not changed if we replace d_i by $d_i' = \text{g.c.d.}(d_i, q-1)$ for $i = 1,2,\dots,n$. We therefore assume, without loss of generality, that $d_i \mid (q-1)$ for $i = 1,2,\dots,n$.

Now consider the equation

$$a_1 x_1^{d_1} + a_2 x_2^{d_2} + \dots + a_n x_n^{d_n} = a_0 , \tag{5.2}$$

admitting, for the moment, any coefficients $(a_0, a_1, \dots, a_n) \in F_q^{n+1}$. Let $N(a_0, a_1, \dots, a_n)$ denote the number of solutions of (5.2) in F_q^n. Then, interchanging sums again, we have

$$\sum_{(a_0,\dots,a_n) \in F_q^{n+1}} N(a_0,\dots,a_n) = \sum_{\underline{x} \in F_q^n} \sum_{\substack{(a_0,\dots,a_n) \\ \text{satisfying (5.2)}}} 1 = \sum_{\underline{x} \in F_q^n} q^n = q^{2n} . \tag{5.3}$$

Thus the mean value of $N(a_0,\dots,a_n)$ is q^{n-1}.

LEMMA 5B:

$$\sum_{(a_0,\dots,a_n) \in F_q^{n+1}} (N(a_0,\dots,a_n) - q^{n-1})^2 \le q^{2n-1}(q-1)\, d_1 d_2 \cdots d_n .$$

Proof:

$$\sum_{(a_0,\ldots,a_n)\in F_q^{n+1}} N^2(a_0,\ldots,a_n)$$

$$= \sum_{(a_0,\ldots,a_n)\in F_q^{n+1}} \left(\sum_{\underline{\underline{x}} \text{ with } (5.2)} 1\right)\left(\sum_{\underline{\underline{y}} \text{ with } (5.4)} 1\right)$$

$$= \sum_{\underline{\underline{x}},\underline{\underline{y}}} \sum_{\substack{(a_0,\ldots,a_n)\in F_q^{n+1} \\ \text{with } (5.2) \text{ and } (5.4)}} 1 \ ,$$

where (5.4) is the equation

$$a_1 y_1^{d_1} + a_2 y_2^{d_2} + \ldots + a_n y_n^{d_n} = a_0 \ . \tag{5.4}$$

Now for fixed $\underline{\underline{x}}$ and fixed $\underline{\underline{y}}$, the system of the two equations (5.2) and (5.4) is a system of two linear homogeneous equations in $a_0,a_1,\ldots,a_n$. This system can have rank 1 or 2 . If the rank is 1 , the number of solutions in $(a_0,\ldots,a_n)$ is q^n . If the rank is 2 , the number of solutions is q^{n-1} . Therefore

$$\sum_{(a_0,\ldots,a_n)\in F_q^{n+1}} N^2(a_0,\ldots,a_n) = \sum_{\underline{\underline{x}},\underline{\underline{y}}} q^{n-1} + \sum_{\substack{\underline{\underline{x}},\underline{\underline{y}} \\ \text{of rank } 1}} (q^n - q^{n-1}) \ .$$

If the matrix

$$\begin{pmatrix} x_1^{d_1} & \ldots & x_n^{d_n} & 1 \\ y_1^{d_1} & \ldots & y_n^{d_n} & 1 \end{pmatrix} \tag{5.5}$$

has rank 1 , then $x_i^{d_i} = y_i^{d_i}$ $(i = 1,2,\ldots,n)$. Since for given $\underline{\underline{x}}$, there are at most d_i possibilities for y_i, hence at most $d_1 \ldots d_n$ possibilities for $\underline{y}$, the number of pairs $\underline{x}, \underline{y}$ such that (5.5) has rank 1 is at most $q^n d_1 d_2 \ldots d_n$. Hence

$$\sum_{(a_0,\ldots,a_n)\in F_q^{n+1}} N^2(a_0,\ldots,a_n) \le q^{3n-1} + q^n d_1 d_2 \ldots d_n (q^n - q^{n-1}) .$$

Using this estimate together with (5.3), we obtain

$$\sum_{(a_0,\ldots,a_n)\in F_q^{n+1}} (N(a_0,\ldots,a_n) - q^{n-1})^2$$

$$= \sum_{(a_0,\ldots,a_n)\in F_q^{n+1}} (N^2(a_0,\ldots,a_n) - q^{2n-2})$$

$$\le q^{3n-1} + q^{2n-1}(q-1) d_1 d_2 \ldots d_n - q^{3n-1}$$

$$= q^{2n-1}(q-1) d_1 d_2 \ldots d_n ,$$

thereby proving Lemma 5B.

To conclude the proof of Theorem 5A, we consider the specific equation (5.1) where $a_1 \neq 0, \ldots, a_n \neq 0$, $a_0 = 1$. Observe that if $t, b_1, b_2, \ldots, b_n$ are non-zero elements of F_q, then

$$N(1, a_1, \ldots, a_n) = N(t, a_1 b_1^{d_1} t, \ldots, a_n b_n^{d_n} t) .$$

The number of distinct $(n+1)$-tuples $(t, b_1^{d_1} t, \ldots, b_n^{d_n} t)$ is

$$(q-1)\left(\frac{q-1}{d_1}\right) \cdots \left(\frac{q-1}{d_n}\right) = \frac{(q-1)^{n+1}}{d_1 d_2 \ldots d_n} ,$$

since $d_i \mid (q-1)$ for $1 \leqq i \leqq n$.

Therefore in the sum of Lemma 5B, there are $(q-1)^{n+1}/(d_1 d_2 \cdots d_n)$ summands which equal $(N - q^{n-1})^2$. So certainly

$$\frac{(q-1)^{n+1}}{d_1 d_2 \cdots d_n}(N - q^{n-1})^2 \le q^{2n-1}(q-1)\, d_1 d_2 \cdots d_n ,$$

and Theorem 5A follows.

THEOREM 5C: *Let* N *be the number of solutions in* F_q^n *of the equation*

$$a_1 x_1^{d_1} + a_2 x_2^{d_2} + \cdots + a_n x_n^{d_n} = 0 , \tag{5.6}$$

where, as above, $(a_1, \ldots, a_n) \in F_q^n$, $a_1 \ne 0$, $a_2 \ne 0$, $\ldots, a_n \ne 0$, *and* $d_i > 0$ $(i = 1,2,\ldots,n)$. *Let* $\delta = \text{l.c.m.}^{\dagger)}\,[d_1, d_2, \ldots, d_n]$. *Then*

$$|N - q^{n-1}| \le \frac{d_1 d_2 \cdots d_n}{\sqrt{\delta}}\, q^{n/2} \left(1 - \frac{1}{q}\right)^{-(n-1)/2} . \tag{5.7}$$

Proof: It is clear that N remains unchanged and that the right hand side of (5.7) cannot increase if we replace d_i by $d_i' = (d_i, q-1)$ and δ by $q' = \text{l.c.m.}\,[d_1', \ldots, d_n']$. Hence we may assume without loss of generality that $d_i \mid (q-1)$ for $1 \leqq i \leqq n$

In the notation used in the proof of Theorem 5A, our $N = N(0, a_1, \ldots, a_n)$. It is clear that

$$N(0, a_1, \ldots, a_n) = N(0, a_1 b_1^{d_1} t, \ldots, a_n b_n^{d_n} t) ,$$

if $t, b_1, \ldots, b_n$ are all non-zero. We need to count the number of

---

†) The least common multiple.

distinct n-tuples $(b_1^{d_1} t, \ldots, b_n^{d_n} t)$. If $(b_1^{d_1} t, \ldots, b_n^{d_n} t) = (b_1'^{d_1} t', \ldots, b_n'^{d_n} t')$, then

$$t'/t = (b_i/b_i')^{d_i} \in (F_q^*)^{d_i} \qquad (i = 1,2,\ldots,n) .$$

Hence $t'/t \in (F_q^*)^{\delta}$, where $\delta = \text{l.c.m.}\,[d_1,\ldots,d_n]$. For given t , there are $(q-1)/\delta$ possibilities for t' in F_q ; and for given t, t', b_i , there are d_i possibilities for b_i' $(i = 1,2,\ldots,n)$. So as $(t, b_1, \ldots, b_n)$ ranges over $F_q^* \times \ldots \times F_q^*$, the number of equal n-tuples is $((q-1)/\delta) d_1 d_2 \ldots d_n$. Thus the number of distinct n-tuples is

$$\frac{(q-1)^{n+1}}{((q-1)/\delta) d_1 d_2 \ldots d_n} = (q-1)^n \frac{\delta}{d_1 d_2 \ldots, d_n} ,$$

and at least that many summands in Lemma 5B are equal to $N - q^{n-1}$. We obtain

$$(q-1)^n \frac{\delta}{d_1 d_2 \ldots d_n} (N - q^{n-1})^2 \le q^{2n-1} (q-1) d_1 d_2 \ldots d_n ,$$

and the theorem follows.

<u>Exercise</u>. Suppose that some exponent d_i in (5.6) is prime to all the others. Then $N = q^{n-1}$.

§6. Additive Equations: Character Sums.

As in Chapter II, we shall write χ for multiplicative characters and ψ for additive characters of F_q.

THEOREM 6A. *Let $f(X_1,\dots,X_n)$ be a polynomial with coefficients in F_q. The number N of zeros of f with coefficients in F_q is given by*

$$N = \frac{1}{q} \sum_{\psi} \sum_{x_1 \in F_q} \cdots \sum_{x_n \in F_q} \psi(f(x_1,\dots,x_n)) , \tag{6.1}$$

where the sum is over additive characters ψ of F_q. This is also given by

$$N = \frac{1}{q} \sum_{a \in F_q} \sum_{x_1 \in F_q} \cdots \sum_{x_n \in F_q} \psi(af(x_1,\dots,x_n)) , \tag{6.2}$$

where $\psi \neq \psi_0$ is a given additive character of F_q.

Proof. The first equation is a consequence of Theorem 1D of Chapter II. Now if $\psi \neq \psi_0$ is fixed, then by Lemma 2D of Chapter II, as a runs through F_q, then $\psi^{(a)}$ with

$$\psi^{(a)}(x) = \psi(ax)$$

runs through all the additive characters. Therefore (6.1) implies (6.2).

THEOREM 6B. *Let N be the number of zeros with components in F_q of*

$$a_1 x_1^{d_1} + \dots + a_n x_n^{d_n} = 0 . \tag{6.3}$$

Suppose $a_i \neq 0$ and $d_i \mid q-1$ $(i = 1,\ldots,n)$. Then if $\psi \neq \psi_0$ is any additive character,

$$N = q^{n-1} + \left(1-\frac{1}{q}\right) \sum_{\substack{\chi_1 \neq \chi_0 \\ \text{of exp } d_1}} \cdots \sum_{\substack{\chi_n \neq \chi_0 \\ \text{of exp } d_n \\ \chi_1 \cdots \chi_n = \chi_0}} \bar{\chi}_1(a_1)\ldots\bar{\chi}_n(a_n) G(\chi_1,\psi)\ldots G(\chi_n,\psi)$$

Here the sum, as indicated, is over certain n-tuples of multiplicative characters, and $G(\chi,\psi)$ denotes Gaussian sums.

Proof. By (6.2) ,

$$\begin{aligned} qN &= \sum_{a \in F_q} \sum_{x_1 \in F_q} \cdots \sum_{x_n \in F_q} \psi(a a_1 x_1^{d_1} + \ldots + a a_n x_n^{d_n}) \\ &= \sum_{a \in F_q} \prod_{i=1}^{n} \left(\sum_{x_i \in F_q} \psi(a a_i x_i^{d_i}) \right) \\ &= q^n + \sum_{a \neq 0} \prod_{i=1}^{n} \left(\sum_{x_i \in F_q} \psi(a a_i x_i^{d_i}) \right) \end{aligned}$$

By Lemma 3B of Chapter II, we have

$$\sum_{x_i} \psi(a a_i x_i^{d_i}) = \sum_{y_i} \psi(a a_i y_i) \sum_{\substack{\chi_i \text{ of} \\ \text{exp } d_i}} \chi_i(y_i) \quad .$$

If $a \neq 0$, we may make the change of variables $y_i \to y_i/(a a_i)$, to obtain

$$\sum_{\substack{\chi_i \text{ of} \\ \exp d_i}} \bar{\chi}_i(a a_i) \sum_{y_i} \chi_i(y_i)\psi(y_i) = \sum_{\substack{\chi_i \text{ of} \\ \exp d_i}} \bar{\chi}_i(a a_i) G(\chi_i, \psi) .$$

Thus

$$qN - q^n = \sum_{\substack{\chi_1 \\ \exp d_1}} \cdots \sum_{\substack{\chi_n \\ \exp d_n}} \bar{\chi}_1(a_1) \ldots \bar{\chi}_n(a_n)$$

$$\left(\sum_{a \neq 0} \bar{\chi}_1(a) \ldots \bar{\chi}_n(a) \right) G(\chi_1, \psi) \ldots G(\chi_n, \psi) .$$

Now if $\chi_1 \ldots \chi_n \neq \chi_0$, then by Theorem 1D of Chapter II,

$$\sum_{a \neq 0} \bar{\chi}_1(a) \ldots \bar{\chi}_n(a) = \sum_a \bar{\chi}_1 \ldots \bar{\chi}_n(a) = 0 .$$

But if $\chi_1 \ldots \chi_n = \chi_0$, then

$$\sum_{a \neq 0} \bar{\chi}_1(a) \ldots \bar{\chi}_n(a) = \sum_{a \neq 0} \chi_0(a) = q - 1 .$$

Moreover, $G(\chi_i, \psi) = 0$ if $\chi_i = \chi_0$ by (3.1) of Chapter II. We therefore may restrict ourselves to $\chi_1, \ldots, \chi_n$ with $\chi_i \neq \chi_0$ $(i = 1, \ldots, n)$ and with $\chi_1 \ldots \chi_n = \chi_0$:

$$qN - q^n = (q - 1) \sum_{\substack{\chi_1 \neq \chi_0 \\ \exp d_1}} \cdots \sum_{\substack{\chi_n \neq \chi_0 \\ \exp d_n \\ \chi_1 \ldots \chi_n = \chi_0}} \bar{\chi}_1(a_1) \ldots \bar{\chi}_n(a_n) G(\chi_1, \psi) \ldots G(\chi_n, \psi) .$$

Theorem 6B follows.

Let g be a fixed generator of the cyclic group F_q^*. A

character χ_i of exponent d_i is of the form $\chi_i(g^t) = e(t\,a_i/d_i)$ $(t = 0,1,2,\ldots)$, where a_i is an integer with $0 \leqq a_i < d_i$. In fact, $0 < a_i < d_i$ if $\chi_i \neq \chi_0$. We have $\chi_1 \cdots \chi_n = \chi_0$ precisely if

$$\frac{a_1}{d_1} + \cdots + \frac{a_n}{d_n} \tag{6.4}$$

is an integer. Thus if $A(d_1,\ldots,d_n)$ is the number of n-tuples of integers $a_1,\ldots,a_n$ with $0 < a_i < d_i$ $(i = 1,\ldots,n)$ and with (6.4) integral, then $A(d_1,\ldots,d_n)$ is also the number of summands in the sum of Theorem 6B. Since the Gaussian sums $G(\chi_i,\psi)$ of Theorem 6B have absolute value $q^{1/2}$ by Theorem 3A of Chapter II, we have

THEOREM 6C. *Make the same hypotheses as in Theorem 6B. Then*

$$|N - q^{n-1}| \leq A(d_1,\ldots,d_n)(1 - \frac{1}{q})q^{n/2} .$$

In particular, $A(d_1,\ldots,d_n) \leq (d_1 - 1) \ldots (d_n - 1)$, so that Theorem 6C is an improvement over Theorem 5C. Theorem 5A could be similarly improved.

Write
$$A_n(d) = A(\underbrace{d,\ldots,d}_{\leftarrow\, n\, \rightarrow})$$

LEMMA 6D. $A_n(d) = \frac{d-1}{d}((d-1)^{n-1} - (-1)^{n-1})$.

Proof. $A_n(d)$ is the number of integers $a_1,\ldots,a_n$ with $0 < a_i < d$ $(i = 1,\ldots,n)$ and $a_1 + \ldots + a_n \equiv 0 \pmod d$. Thus $A_1(d) = 0$ and $A_2(d) = d - 1$, and the formula is correct for $n = 1$ or $n = 2$. For $n \geqq 2$, an n-tuple $a_1,\ldots,a_n$ will be counted by A_n precisely if $0 < a_1 < d, \ldots, 0 < a_{n-1} < d$, $0 < a_n < d$ and

$$-a_n \equiv a_1 + \ldots + a_{n-1} \not\equiv 0 \pmod{d} .$$

The number of possibilities for $a_1, \ldots, a_{n-1}$ is $(d-1)^{n-1} - A_{n-1}(d)$, so that

$$A_n(d) = (d-1)^{n-1} - A_{n-1}(d) .$$

The lemma now follows by induction on n .

COROLLARY 6E. *Suppose* $d \mid q-1$ *and suppose* $a_1, \ldots, a_n$ *are non-zero in* F_q . *The number of* N *of solutions with components in* F_q *of*

$$a_1 x_1^d + \ldots + a_n x_n^d = 0$$

satisfies

$$|N - q^{n-1}| \leq ((d-1)/d)((d-1)^{n-1} - (-1)^{n-1})(1 - q^{-1})q^{n/2} .$$

Following Weil (1949) , we now study the dependence of the number of solutions on the field of coordinates. Again, let $a_1, \ldots, a_n$ be non-zero in F_q , and let $d_i \mid q-1$ $(i = 1, \ldots, n)$. We write N_ν for the number of solutions of (6.3) with coordinates in F_{q^ν} .

If χ_i is a character of F_q of exponent d_i , then $\chi_{\nu i}$ given by

$$\chi_{\nu i}(x) = \chi_i(\mathfrak{N}(x)) ,$$

where $\mathfrak{N}$ is the norm $F_{q^\nu} \to F_q$, is a character of F_{q^ν} of exponent d_i. Since $\mathfrak{N}$ is onto F_q, it follows that $\chi_{\nu i} \neq \chi'_{\nu i}$ if $\chi_i \neq \chi'_i$. Therefore as χ_i runs through all the characters of F_q of exponent d_i, then $\chi_{\nu i}$ runs through all the characters of F_{q^ν} of exponent d_i. Moreover, we may replace the character ψ in Theorem 6B by ψ_ν with

$$\psi_\nu(x) = \psi(\mathfrak{T}(x)) \quad ,$$

where $\mathfrak{T}$ is the trace $F_{q^\nu} \to F_q$. In the formula of Theorem 6B, we have to replace q by q^ν, $\bar{\chi}_i(a_i)$ by $\bar{\chi}_{\nu i}(a_i) = (\bar{\chi}_i(a_i))^\nu$, and $G = G(\chi_i,\psi)$ by $G_\nu = G(\chi_{\nu i},\psi_\nu)$, which by the Davenport-Hasse Relation (Corollary 10E of Chapter II) has

$$-G_\nu = (-G)^\nu \quad .$$

Thus

$$N_\nu = (q^{n-1})^\nu + (-1)^{n(\nu-1)}\left(1 - \frac{1}{q^\nu}\right) \sum_{\substack{\chi_1 \neq \chi_0 \\ \exp d_1}} \cdots \sum_{\substack{\chi_n \neq \chi_0 \\ \exp d_n \\ \chi_1 \cdots \chi_n = \chi_0}} (\bar{\chi}_1(a_1) \cdots \bar{\chi}_n(a_n) G(\chi_1,\psi) \cdots G(\chi_n,\psi))^\nu .$$

Thus N_ν is of the form

$$N_\nu = \omega_1^\nu + \dots + \omega_u^\nu - \eta_1^\nu - \dots - \eta_v^\nu , \tag{6.5}$$

where $\omega_1,\dots,\omega_u$, $\eta_1,\dots,\eta_v$ are complex algebraic numbers, with absolute values

$$|\omega_i| = q^{c_i/2} \; , \; |\eta_j| = q^{d_j/2} \; , \tag{6.6}$$

where the c_i and d_j are integers.

Weil (1949) made the famous conjecture that a formula like (6.5) with (6.6) is true in general for the number N_ν of solutions in $F^n_{q^\nu}$ of $f(x_1,\ldots,x_n) = 0$, where f is a polynomial with coefficients in F_q . In fact, the conjecture said much more than this.

For curves, i.e. for $n = 2$, the truth of this follows from the Riemann Hypothesis for curves, which had been proved by Weil (1940, 1948a). It may be deduced from Theorem 1A of Chapter III and the theory of the Zeta Function of a curve (Artin(1924), F.K. Schmidt (1931). A very readable text is Deuring (1958)). For general n , the part (6.5) of the conjecture was first proved by Dwork (1960). The general conjecture was proved by Deligne (1973).+)

+) But see the remark in the Preface.

§7. Equations $f_1(y)x_1^{d_1} + \dots + f_n(y)x_n^{d_n} = 0$.

THEOREM 7A. *Let $f_1(Y),\dots,f_n(Y)$ be non-zero polynomials of degree $\le m$ over F_q. Suppose they are coprime in pairs. Further suppose that if $a_1,\dots a_n$ are integers with*

(7.1) $\quad 0 < a_j < d_j \quad (j = 1,\dots,n)$ *and* $\frac{a_1}{d_1} + \dots + \frac{a_n}{d_n}$ *integral* ,

and if $\delta = \text{l.c.m.}(d_1,\dots,d_n)$, *then the polynomial*

(7.2) $$f_1(Y)^{a_1\delta/d_1} \cdots f_n(Y)^{a_n\delta/d_n}$$

is not a δ th *power. Then the number* N *of solutions* $x_1,\dots,x_n,y$ *with components in* F_q *of the equation in the title satisfies*

$$|N - q^n| \le c_1(n,m,\delta)q^{(n+1)/2}$$

A special case is when the polynomials f_j are coprime and if n is odd and $d_1 = \dots = d_n = 2$. For then there exist no integers $a_1,\dots,a_n$ with (7.1), and the hypothesis is satisfied.

Another special case is when the f_j are coprime and there is an i in $1 \le i \le n$ such that $f_i(Y) - X^{d_i}$ is absolutely irreducible. For then the polynomials (7.2) are not δ th powers: To see this, it will suffice, because of the coprime condition, that $f_i(Y)^{a_i\delta/d_i}$ is not a δ th power, which is the same as that $f_i(Y)^{a_i}$ is not a d_i th power. Now if $f_i(Y) = c(Y-\alpha_1)^{c_1} \dots (Y-\alpha_s)^{c_s}$, then $(d_i, c_1,\dots,c_s) = 1$ by Lemma 2C of Ch. I, so that $(d_i, a_ic_1,\dots,a_ic_s) \le a_i < d_i$, and indeed $f_i(Y)^{a_i}$ is not a d_ith power.

Proof. We shall write $g(q) = O(h(q))$ if $|g(q)| \leq c(n,m,\delta)h(q)$. Thus the assertion of the theorem is that $N = q^n + O(q^{(n+1)/2})$. As before (see, e.g., §5 of Ch. II), we may reduce the proof to the case when $d_j | q - 1$ $(j = 1,\ldots,n)$.

Suppose $y \in F_q$ has $f_1(y) = 0$. Then $f_2(y)\ldots f_n(y) \neq 0$. The number of $x_2,\ldots,x_n$ with $f_2(y)x_2^{d_2} + \ldots + f_n(y)x_n^{d_n} = 0$ is $q^{n-2} + O(q^{(n-1)/2})$ by Theorem 5C or 6C. Since there are q possibilities for x_1, we obtain $q^{n-1} + O(q^{(n+1)/2})$ solutions with this particular value of y. The number N_1 of solutions of the equation of the title with

$$f_1(y) \ldots f_n(y) = 0 \tag{7.3}$$

is therefore

$$N_1 = Zq^{n-1} + O(q^{(n+1)/2}) ,$$

where Z is the number of y in F_q with (7.3).

For given y with

$$f_1(y) \ldots f_n(y) \neq 0 , \tag{7.4}$$

the number of solutions of our equation in $x_1,\ldots,x_n$ is given by Theorem 6B. Therefore the number N_2 of solutions of our equation with (7.4) is

$$N_2 = q^{n-1}(q - Z) + R$$

where

$$|R| \leq \sum_{\substack{\chi_1 \neq \chi_0 \\ \exp d_1}} \cdots \sum_{\substack{\chi_n \neq \chi_0 \\ \exp d_n \\ \chi_1 \cdots \chi_n = \chi_0}} \left| \sum_{\substack{y \text{ with} \\ (7.4)}} \bar{\chi}_1(f_1(y)) \cdots \bar{\chi}_n(f_n(y)) \right| q^{n/2} ,$$

since $|G(\chi_i, \psi)| = q^{1/2}$.

Let χ be a character of order δ . Then $\chi_j = \chi^{a_j \delta / d_j}$ for some a_j in $0 < a_j < d_j$ $(j = 1, \ldots, n)$. Since $\chi_1 \cdots \chi_n = \chi_0$, the conditions (7.1) hold, and (7.2) is not a δ th power. The inner sum in our estimate of $|R|$ is[†)]

$$\sum_y \bar{\chi}(f_1(y)^{a_1 \delta / d_1} \cdots f_n(y)^{a_n \delta / d_n}) ,$$

and by Theorem 2B′ of Ch. II, it is $O(q^{1/2})$. Thus $R = O(q^{(n+1)/2})$, and

$$N = N_1 + N_2 = q^n + R + O(q^{(n+1)/2}) = q^n + O(q^{(n+1)/2}).$$

THEOREM 7B. *Let N be the number of solutions of*

$$f_1(y) x_1^{d_1} + \cdots + f_{n-1}(y) x_{n-1}^{d_{n-1}} + f_n(y) = 0 \tag{7.5}$$

in $x_1, \ldots, x_{n-1}, y$ *in* F_q . *Put* $d_n = \text{l.c.m.}(d_1, \ldots, d_{n-1})$, *and suppose that* $f_1, \ldots, f_n$ *satisfy the conditions of Theorem 7A. Then*

$$|N - q^{n-1}| \leq c_2(n, m, d_n) q^{(n-1)/2} .$$

[†)] The condition (7.4) in the sum is immaterial.

This generalizes a result of Perelmuter and Postnikov (1972).

Proof. Let $\tilde{N}$ be the number of solutions in $\tilde{x}_1,\ldots,\tilde{x}_n,\tilde{y}$ of

$$f_1(\tilde{y})\tilde{x}_1^{d_1}+\ldots+f_{n-1}(\tilde{y})\tilde{x}_{n-1}^{d_{n-1}}+f_n(\tilde{y})\tilde{x}_n^{d_n}=0 \quad . \tag{7.6}$$

If $x_1,\ldots,x_{n-1},y$ is a solution of (7.5), then for $x \neq 0$,

$$\tilde{x}_1 = x_1 x^{d_n/d_1},\ldots,\tilde{x}_{n-1} = x_{n-1}x^{d_n/d_{n-1}},\tilde{x}_n = x,\tilde{y} = y$$

is a solution of (7.6) with $\tilde{x}_n \neq 0$. Every solution of (7.6) with $\tilde{x}_n \neq 0$ is obtained in this way. The solutions of (7.6) with $\tilde{x}_n = 0$ number $q^{n-1} + O(q^{(n+1)/2})$, as is seen as follows:

If $f_1(\tilde{y}) = 0$, then $f_2(\tilde{y}) \ldots f_{n-1}(\tilde{y}) \neq 0$, and the number of $\tilde{x}_2,\ldots,\tilde{x}_{n-1}$ with $f_2(\tilde{y})\tilde{x}_2^{d_2}+\ldots+f_{n-1}(\tilde{y})\tilde{x}_{n-1}^{d_{n-1}} = 0$ is $q^{n-3} + O(q^{(n-2)/2})$ by Theorem 5C or 6C. Thus the number of solutions of (7.6) with $\tilde{x}_n = 0$ with $f_1(\tilde{y}) \ldots f_{n-1}(\tilde{y}) = 0$ is $Zq^{n-2} + O(q^{n/2})$ where Z is the number of $\tilde{y}$ with $f_1(\tilde{y})\ldots f_{n-1}(\tilde{y}) = 0$. On the other hand, the number of solutions of (7.6) with $\tilde{x}_n = 0$ and $f_1(\tilde{y}) \ldots f_{n-1}(\tilde{y}) \neq 0$ is $(q - Z)(q^{n-2} + O(q^{(n-1)/2}))$, again by Theorem 5C or 6C. Together we get $Zq^{n-2} + (q - Z)q^{n-2} + O(q^{(n+1)/2}) = q^{n-1} + O(q^{(n+1)/2})$.

Thus

$$\tilde{N} = (q - 1)N + q^{n-1} + O(q^{(n+1)/2}) \quad .$$

Now $\tilde{N} = q^n + O(q^{(n+1)/2})$ by Theorem 7A, and therefore

$$(q - 1)N = (q - 1)q^{n-1} + O(q^{(n+1)/2}) \quad .$$

V. Absolutely Irreducible Equations $f(x_1,\ldots,x_n) = 0$.

References: Ostrowski (1919), Noether (1922), Lang & Weil (1954), Nisnevich (1954).

§1. Elimination theory.

Our goal is to derive an estimate for the number of zeros of an absolutely irreducible polynomial in n variables. This will be achieved in §5 . But in order to reach this goal we need "Bertini's Theorem", and for that in turn we need elimination theory. For more information on elimination theory see Van der Waerden (1955) , Chapter 11 . Elimination theory is now considered old fashioned, since most of its applications can be derived in a more elegant way from algebraic geometry. On the other hand, in these lectures we do not presume any knowledge of algebraic geometry. Moreover, elimination theory is constructive and easily permits one to estimate the degrees and the size of the coefficients of the constructed polynomials.

The reader will recall that given two polynomials over a field K ,

$$f(X) = c_0X^a + c_1X^{a-1} + \ldots + c_a ,$$

$$g(X) = d_0X^b + d_1X^{b-1} + \ldots + d_b ,$$

the resultant $R = R(c_0,c_1,\ldots,c_a,d_0,d_1,\ldots,d_b)$ of $f(X)$ and $g(X)$ is a certain polynomial in the coefficients of f and g . The polynomial R vanishes precisely if either f and g have a common root or if both leading coefficients are zero ($c_0 = d_0 = 0$) . If $c_0 \neq 0$ and $d_0 \neq 0$, then

$$R = c_0^b d_0^a \prod_{i=1}^{a} \prod_{j=1}^{b} (y_i - z_j) ,$$

where $y_1,\ldots,y_a$ and $z_1,\ldots,z_b$ are the roots of f and of g , respectively. R is homogeneous of degree b in $c_0,\ldots,c_a$, and homogeneous of degree a in $d_0,\ldots,d_b$, and each term $c_0^{i_0} c_1^{i_1} \ldots c_a^{i_a} d_0^{j_0} d_1^{j_1} \ldots d_b^{j_b}$ has

$$(i_1 + 2i_2 + \ldots + ai_a) + (j_1 + 2j_2 + \ldots + bj_b) = ab .$$

Let

$$f^*(X_0,X_1) = c_0X_1^a + c_1X_0X_1^{a-1} + \ldots + c_aX_0^a$$

and

$$g^*(X_0,X_1) = d_0X_1^b + d_1X_0X_1^{b-1} + \ldots + d_bX_0^b$$

be the two forms associated with $f(X)$ and $g(X)$. We say that a pair (x_0,x_1) is a <u>common zero</u> of f^* and g^* if $(x_0,x_1) \neq (0,0)$ and $f^*(x_0,x_1) = g^*(x_0,x_1) = 0$, and if $x_0,x_1 \in \bar{K}$.

<u>Claim:</u> $f^*(X_0,X_1)$ <u>and</u> $g^*(X_0,X_1)$ <u>have a common zero if and only if</u> $R = 0$.

<u>Proof:</u> First suppose that f^* and g^* have the common zero (x_0,x_1) . If $x_0 \neq 0$ then they have a common zero of the form $(1,z)$. Here z is a common root of f and g , and therefore $R = 0$. If $x_0 = 0$, then $c_0x_1^a = 0$ and $d_0x_1^b = 0$. Since x_1 cannot also be zero, it follows that $c_0 = d_0 = 0$, and $R = 0$.

Now suppose $R = 0$. Either f and g have a common root z , in which case f^* and g^* have the common root $(1,z)$. Or $c_0 = d_0 = 0$, in which case $(1,0)$ is a common root of f^* and g^* . This verifies the claim. It follows that the vanishing of the resultant has a more elegant interpretation in terms of f^* and g^* than of f and g .

Let $f_1(X_0,X_1,\ldots,X_k)$, ... , $f_r(X_0,X_1,\ldots,X_k)$ be forms with coefficients in a field K . A <u>common zero</u> of $f_1,\ldots,f_r$ is an $(n+1)$-tuple $(x_0,x_1,\ldots,x_k) \neq \underline{\underline{0}}$ with components in $\bar{K}$ such that $f_i(\underline{\underline{x}}) = 0$ for $i = 1,2,\ldots,r$. Suppose each of these forms is of degree d , and that for $j = 1,2,\ldots,r$,

$$(1.1)\quad f_j(X_0,X_1,\ldots,X_k) = \sum_{i_0+i_1+\ldots+i_k=d} a^{(j)}_{i_0i_1\ldots i_k} X_0^{i_0} X_1^{i_1} \ldots X_k^{i_k} .$$

We extend the concept of a resultant of two polynomials to a resultant system for r forms in $k+1$ variables by giving the following

<u>Definition</u>: A <u>resultant system</u> for the forms (1.1) is a finite set of forms $g_1,\ldots,g_s$ in variables

$$A^{(j)}_{i_0i_1\ldots i_k} \qquad (1 \leq j \leq r\ ;\ i_0 + i_1 + \ldots + i_k = d) ,$$

with the property that $g_i\left(a^{(j)}_{i_0i_1\ldots i_k}\right) = 0$ for each $i = 1,\ldots,s$ if and only if the forms $f_1,\ldots,f_r$ have a common zero.

<u>Example 1</u>: Take $k = 1$ and $r = 2$. The resultant system for the forms $f_1(X_0,X_1)$, $f_2(X_0,X_1)$ consists of just one form $(s = 1)$ - the resultant of the two polynomials $f_1(1,X_1)$ and $f_2(1,X_1)$.

<u>Example 2</u>: Take

$$f_1(X_1,\ldots,X_n) = a_{11}X_1 + \ldots + a_{1n}X_n ,$$
$$\vdots$$
$$f_n(X_1,\ldots,X_n) = a_{n1}X_1 + \ldots + a_{nn}X_n ,$$

i.e. a set of n linear forms in n variables. Again there is a resultant system for these forms consisting of a single form g, namely the determinant

$$g = \begin{vmatrix} A_{11} & A_{12} & \cdots & A_{1n} \\ \vdots & \vdots & & \vdots \\ A_{n1} & A_{n2} & \cdots & A_{nn} \end{vmatrix} .$$

More generally, we can describe a resultant system for the forms

$$\begin{aligned} f_1(X_1,\ldots,X_n) &= a_{11}X_1 + \cdots + a_{1n}X_n \\ &\vdots \\ f_m(X_1,\ldots,X_n) &= a_{m1}X_1 + \cdots + a_{mn}X_n . \end{aligned}$$

If $m < n$, a resultant system for the forms $f_1,\ldots,f_m$ is the identically zero form, since $f_1,\ldots,f_m$ always have a common zero. If $m \geq n$, a resultant system is the set of all $(n \times n)$- subdeterminants of the associated $m \times n$ matrix.

THEOREM 1A: *Let $f_1(X_0,X_1,\ldots,X_k)$, ..., $f_r(X_0,X_1,\ldots,X_k)$ be forms of degree d as in (1.1). There exists a resultant system $g_1,\ldots,g_s$, where each g_i is a form in the variables $A^{(j)}_{i_0 i_1 \ldots i_k}$ of degree*

$$2^k d^{2^k - 1} .$$

LEMMA 1B: *Let $\hat{g}(X_1,\ldots,X_m)$ be a form of degree e, and let $h_1(Y_1,\ldots,Y_\ell),\ldots,h_m(Y_1,\ldots,Y_\ell)$ be forms of degree e'. Then the polynomial*

$$g(Y_1,\dots,Y_\ell) = \hat{g}(h_1(Y_1,\dots,Y_\ell),\dots,h_m(Y_1,\dots,Y_\ell))$$

<u>is a form of degree</u> ee'.

<u>Proof</u>: Obvious.

We begin the

<u>Proof of Theorem 1A</u>: Let the forms $f_1(X_0,\dots,X_k),\dots,f_r(X_0,\dots,X_k)$ be given by (1.1). The proof is by induction on k. If $k = 0$, then (1.1) becomes

(1.2) $$f_j(X_0) = a_d^{(j)} X_0^d \quad , \qquad 1 \le j \le r .$$

Clearly the forms

$$g_j(A_d^{(1)},\dots,A_d^{(r)}) = A_d^{(j)} \quad , \qquad 1 \le j \le r ,$$

form a resultant system for (1.2). Moreover, $\deg g_j(A_d^{(1)},\dots,A_d^{(r)}) = 1$ for $1 \le j \le r$, which agrees with Theorem 1A.

Suppose that the theorem holds for forms in k variables $X_0,X_1,\dots,X_{k-1}$. We introduce new variables $U_1,\dots,U_r$, $V_1,\dots,V_r$, and form two polynomials

$$\bar{f} = U_1 f_1(X_0,\dots,X_k) + \dots + U_r f_r(X_0,\dots,X_k) ,$$

$$\bar{g} = V_1 f_1(X_0,\dots,X_k) + \dots + V_r f_r(X_0,\dots,X_k) ,$$

where

$$f_j(X_0,\dots,X_k) = \sum_{i_0+\dots+i_k=d} A^{(j)}_{i_0\dots i_k} X_0^{i_0}\dots X_k^{i_k} .$$

If we view $\bar{f}$ and $\bar{g}$ as polynomials in the variable X_k, they have a resultant

$$R = R(X_0,\ldots,X_{k-1},U_1,\ldots,U_r,V_1,\ldots,V_r,\text{all } A's)^{\dagger)} .$$

If we write

$$\bar{f} = \bar{a}_0 X_k^d + \ldots + \bar{a}_d$$

and

$$\bar{g} = \bar{b}_0 X_k^d + \ldots + \bar{b}_d ,$$

then each $\bar{a}_i$ and each $\bar{b}_i$ is a form of degree i in $X_0,\ldots,X_{k-1}$, is linear in the variables $U_1,\ldots,U_r,V_1,\ldots,V_r$, and linear in the $A's$.

In the resultant, a term $\bar{a}_0^{j_0} \ldots \bar{a}_d^{j_d} \bar{b}^{\ell_0} \ldots \bar{b}_d^{\ell_d}$ has

$$j_1 + 2j_2 + \ldots + dj_d + \ell_1 + 2\ell_2 + \ldots + d\ell_d = d^2 .$$

The resultant is of degree d in $\bar{a}_0,\ldots,\bar{a}_d$, and also of degree d in $\bar{b}_0,\ldots,\bar{b}_d$. Therefore

(i) R is a form of degree d^2 in $X_0,\ldots,X_{k-1}$;

(ii) R is a form of degree $2d$ in the $A's$;

(iii) R is a form of degree $2d$ in $U_1,\ldots,U_r$, $V_1,\ldots,V_r$ together.

Collecting terms involving like powers in the $U's$ and $V's$, we may certainly write

$$R = \sum_{u_1,\ldots,u_r} \sum_{v_1,\ldots,v_r} R_{u_1,\ldots,u_r,v_1,\ldots,v_r}(X_0,\ldots,X_{k-1},A's)\, U_1^{u_1}\ldots U_r^{u_r} V_1^{v_1}\ldots V_r^{v_r} .$$

---

†) That is, all variables A.

Abbreviating the above coefficients by $R_{\underline{\underline{u}},\underline{\underline{v}}}$, we observe that

(i) $R_{\underline{\underline{u}},\underline{\underline{v}}}$ is a form of degree d^2 in $X_0,\ldots,X_{k-1}$;

(ii) $R_{\underline{\underline{u}},\underline{\underline{v}}}$ is a form of degree $2d$ in the $A's$.

LEMMA 1C: Suppose the variables $A^{(j)}_{i_0,\ldots i_k}$ are replaced by coefficients $a^{(j)}_{i_0\ldots i_k}$ in the field K . Then $f_1,\ldots,f_r$ have a common zero if and only if all of the polynomials $R_{\underline{\underline{u}},\underline{\underline{v}}}(X_0,\ldots,X_{k-1},a's)$ have a common zero.

Proof: Suppose $f_1,\ldots,f_r$ have a common zero $(x_0,x_1,\ldots,x_k)$. If $(x_0,x_1,\ldots,x_{k-1}) \neq (0,0,\ldots,0)$ and the values $x_0,x_1,\ldots,x_{k-1}$ are substituted in $\bar{f}$ and $\bar{g}$, then x_k is a common zero of $\bar{f}$ and $\bar{g}$, whence $R = 0$. But since

$$0 = R = \sum_{\underline{\underline{u}}} \sum_{\underline{\underline{v}}} R_{\underline{\underline{u}},\underline{\underline{v}}}(x_0,\ldots,x_{k-1},a's)\, U_1^{u_1}\ldots U_r^{u_r} V_1^{v_1}\ldots V_r^{v_r} ,$$

the polynomials $R_{\underline{\underline{u}},\underline{\underline{v}}}(X_0,\ldots,X_{k-1},a's)$ must have $(x_0,\ldots,x_{k-1})$ as a common zero. If, on the other hand, $(x_0,\ldots,x_{k-1}) = (0,\ldots,0)$, then $f_1,\ldots,f_r$ have the common zero $(0,\ldots,0,1)$. It follows that the coefficient of X_k^d is zero for each f_i , hence also for $\bar{f}$ and $\bar{g}$. Again $R = R(X_0,\ldots,X_{k-1},U's,V's,a's) = 0$, so all of the forms $R_{\underline{\underline{u}},\underline{\underline{v}}}(X_0,\ldots,X_{k-1},a's)$ are identically zero, and therefore have a non-trivial common zero.

Conversely, suppose that $(x_0,x_1,\ldots,x_{k-1})$ is a common zero of the forms $R_{\underline{\underline{u}},\underline{\underline{v}}}(X_0,\ldots,X_{k-1},a's)$. In particular, $x_0,\ldots,x_{k-1}$ lie in $\bar{K}$. Then

$$R(x_0,x_1,\ldots,x_{k-1},U's,V's,a's) = 0 ,$$

so that either $\bar{a}_0 = \bar{b}_0 = 0$ or $\bar{f}$ and $\bar{g}$ have a common zero x_k. If $\bar{a}_0 = \bar{b}_0 = 0$, then $f_1,\ldots,f_r$ clearly have the common zero $(0,0,\ldots,0,1)$. If $\bar{f}$ and $\bar{g}$ have the common root x_k, then x_k as a root of $\bar{f}$ is algebraic over $K(U_1,\ldots,U_r)$, and as a root of $\bar{g}$ is algebraic over $K(V_1,\ldots,V_r)$. It follows that x_k is algebraic over K. But since

$$\bar{f} = U_1f_1(x_0,\ldots,x_k) + \ldots + U_rf_r(x_0,\ldots,x_k) = 0 ,$$

and since each $f_j(x_0,\ldots,x_k)\in\bar{K}$, we conclude that

$$f_j(x_0,\ldots,x_k) = 0 \qquad (1 \le j \le r) .$$

We now return to the proof of Theorem 1A. By the inductive hypothesis, there is a resultant system $\hat{g}_1,\ldots,\hat{g}_s$ for the forms $R_{\underline{\underline{u}},\underline{\underline{v}}}(X_0,\ldots,X_{k-1})$, with

$$\deg \hat{g}_i = 2^{k-1}(d^2)^{2^{k-1}-1} = 2^{k-1}d^{2^k-2} \qquad (1 \le i \le s) .$$

Each coefficient of $R_{\underline{\underline{u}},\underline{\underline{v}}}$ was a form of degree $2d$ in the $A's$. Let $g_1,\ldots,g_s$ be obtained from $\hat{g}_1,\ldots,\hat{g}_s$ by substituting for each coefficient of $R_{\underline{\underline{u}},\underline{\underline{v}}}$ its expression in terms of the $A's$. By Lemma 1C, it is obvious that $g_1,\ldots,g_s$ form a resultant system for $f_1,\ldots,f_r$. Finally, by Lemma 1B, each g_i is a form in the $A's$ of degree

$$2d \cdot 2^{k-1} d^{2^k - 2} = 2^k d^{2^k - 1} \quad .$$

This concludes the proof of Theorem 1A. We remark that the forms $g_1, \ldots, g_s$ have rational integer coefficients and are independent of the field K if char $K = 0$. In a field of characteristic p, the coefficients of the forms $g_1, \ldots, g_s$ are replaced by the residue classes modulo p of the corresponding coefficients in characteristic zero.

If a is a polynomial with rational integer coefficients in any number of variables, we define $\|a\|$ as the sum of the absolute values of the coefficients. For

Example: If $a(X,Y) = (X-Y)^n$, then $\|a\| = 2^n$.

Theorem 1D: In a field of characteristic zero, the forms $g_1, \ldots, g_s$ of Theorem 1A have rational integer coefficients and satisfy

$$\|g_i\| \leq 2^{2^{4k}} \cdot d^{2^k} \qquad (1 \leq i \leq s) \; .$$

For the remainder of this section, all polynomials are assumed to have rational integer coefficients. We first prove an analog to Lemma 1B.

LEMMA 1E: Let $\hat{g}(X_1, \ldots, X_m)$ be a polynomial of total degree e. Let $b_1(Y_1, \ldots, Y_t), \ldots, b_m(Y_1, \ldots, Y_t)$ be polynomials with $\|b_i\| \leq \psi$ $(1 \leq i \leq m)$. Then

$$g(Y_1, \ldots, Y_t) = \hat{g}(b_1(Y_1, \ldots, Y_t), \ldots, b_m(Y_1, \ldots, Y_t))$$

has the property that

$$\|g\| \le \|\hat{g}\| \psi^{e} .$$

Proof: For any two polynomials a and b in any number of variables, observe that

$$\|ab\| \le \|a\| . \|b\| .$$

For if a', b' and $(ab)'$ are obtained from a , b and ab , respectively, by replacing each coefficient by its absolute value, then

$$\|ab\| = \|(ab)'\| \le \|a'b'\| = \|a'\| \ \|b'\| = \|a\| \ \|b\| .$$

Now a typical term in the polynomial g is $b_1^{i_1} \ldots b_m^{i_m}$ where

$$i_1 + i_2 + \ldots + i_m \le e ,$$

so that

$$\|b_1^{i_1} \ldots b_m^{i_m}\| \le \psi^{e} .$$

The lemma follows.

In order to prove Theorem 1D , we examine more closely the polynomials introduced in the proof of Theorem 1A .

LEMMA 1F:

$$\|R_{\underline{\underline{u}},\underline{\underline{v}}}(X_0,\ldots,X_{k-1},A's)\| \le (2d)^{6dk} .$$

Proof: We saw that $R(X_0,\ldots,X_{k-1},U_1,\ldots,U_r,V_1,\ldots,V_r,A's)$ had total degree $2d$ in $U_1,\ldots,U_r,V_1,\ldots,V_r$. Therefore in each monomial $U_1^{u_1} \ldots U_r^{u_r} V_1^{v_1} \ldots V_r^{v_r}$,

$$u_1 + \dots + u_r + v_1 + \dots + v_r \le 2d .$$

Hence for any $R_{\underline{\underline{u}},\underline{\underline{v}}}$ which is not identically zero, at most $2d$ of the numbers $u_1,\dots,u_r,v_1,\dots,v_r$ can be non-zero. Let $c = \min\{2d,r\}$. Suppose, without loss of generality, that $u_i = v_i = 0$ if $i > c$. Let $R^*(X_0,\dots,X_{k-1},U_1,\dots,U_c,V_1,\dots,V_c,A's)$ be obtained from R by omitting all terms where some U_i or V_i with $i > c$ occurs. Then

$$\|R_{\underline{\underline{u}},\underline{\underline{v}}}(X_0,\dots,X_{k-1},A's)\| \le \|R^*\| .$$

R^* is clearly the resultant of the two polynomials

$$\bar{f}^* = U_1 f_1(X_0,\dots,X_k) + \dots + U_c f_c(X_0,\dots,X_k)$$

$$= \bar{a}_0^* X_k^d + \dots + \bar{a}_d^* ,$$

$$\bar{g}^* = V_1 f_1(X_0,\dots,X_k) + \dots + V_c f_c(X_0,\dots,X_k)$$

$$= \bar{b}_0^* X_k^d + \dots + \bar{b}_d^* ,$$

when considered as polynomials in X_k. If we write, for $1 \le j \le r$,

$$f_j(X_0,\dots,X_k) = \sum_{i_0+\dots+i_k=d} A^{(j)}_{i_0\dots i_k} X_0^{i_0}\dots X_k^{i_k} ,$$

the number of summands in f_j is not more than $(d+1)^k$. So the number of summands in $\bar{f}^*$ or $\bar{g}^*$ is bounded by

$$(d+1)^k c \le 2d(d+1)^k \le (2d)^{k+1} .$$

Therefore the number of summands in each $\bar{a}_i^*$ or $\bar{b}_i^*$ is also bounded by $(2d)^{k+1}$. But each coefficient in $\bar{a}_i^*$ or $\bar{b}_i^*$ is either 0 or 1, so that

$$\|\bar{a}_i^*\| \leq (2d)^{k+1} , \quad \|\bar{b}_i^*\| \leq (2d)^{k+1} \qquad (i = 0,\dots,d) .$$

The resultant of $\bar{f}^*$ and $\bar{g}^*$ is of degree $2d$ in $\bar{a}_0^*,\dots,\bar{a}_d^*,\bar{b}_0^*,\dots,\bar{b}_d^*$. This resultant is a $(2d \times 2d)$ - determinant, so the resultant r satisfies $\|r\| \leq (2d)!$. By Lemma 1E, $R^* = r(a_0^*,\dots,a_d^*, b_0^*,\dots,b_d^*)$ has

$$\|R^*\| \leq (2d)!\ ((2d)^{k+1})^{2d} .$$

Hence

$$\begin{aligned}\|R_{\underline{\underline{u}},\underline{\underline{v}}}(X_0,\dots,X_{k-1},A-s)\| &\leq \|R^*\| \\ &\leq (2d)^{2d}(2d)^{2dk+2d} \\ &= (2d)^{2dk+4d} \\ &\leq (2d)^{6dk} .\end{aligned}$$

<u>Proof of Theorem 1D:</u> We proceed by induction on k. If $k = 0$, then $\|g_i\| = 1$ and the theorem holds trivially. Suppose it has been established that for $k-1$ one obtains the estimate

$$\|g_i\| \leq c_{k-1} = c_{k-1}(d) = 2^{2^{4(k-1)}\cdot d^{2^{k-1}}} .$$

Let $\hat{g}_1,\dots,\hat{g}_s$ be a resultant system for the $R_{\underline{\underline{u}},\underline{\underline{v}}}$'s. By induction, $\|\hat{g}_i\| \leq c_{k-1}(d^2)$, since each $R_{\underline{\underline{u}},\underline{\underline{v}}}$ is of degree d^2 in $X_0,\dots,X_{k-1}$.

On the other hand, g_i is obtained from $\hat{g}_i$ by substituting for the coefficients of each $R_{\underline{\underline{u}},\underline{\underline{v}}}$ their expressions in terms of the A's . By applying Lemmas 1E , 1F and observing that $\hat{g}_i$ has degree

$$2^{k-1}(d^2)^{2^{k-1}-1} = 2^{k-1}d^{2^k-2} \quad ,$$

we obtain

$$\|g_i\| \le c_{k-1}(d^2) \left|(2d)^{6kd}\right|^{2^{k-1}d^{2^k-2}} \quad .$$

But by the inductive hypothesis,

$$c_{k-1}(d^2) = 2^{2^{4k-4}d^{2^k}} \quad .$$

Hence

$$\|g_i\| \le 2^{2^{4k-4}d^{2^k}} \cdot 2^{2d \cdot 6kd \cdot 2^{k-1} \cdot d^{2^k-2}}$$

$$= 2^{2^{4k-4}d^{2^k}} \cdot 2^{6k\,2^k d^{2^k}}$$

$$= 2^{(2^{4k-4} + 6k2^k)d^{2^k}}$$

$$< 2^{2^{4k}d^{2^k}} \quad .$$

§2. The absolute irreducibility of polynomials (I) .

Given a polynomial $f(X_1,\dots,X_n)$ in n variables with coefficients in a field K , we wish to investigate the absolute irreducibility of f ; i.e., the irreducibility of f over $\bar{K}$. Suppose f has total degree at most $d > 0$ and is given by

$$(2.1)\qquad f(X_1,\dots,X_n) = \sum_{i_1+\dots+i_n \le d} a_{i_1\dots i_n} X_1^{i_1}\dots X_n^{i_n} .$$

THEOREM 2 A: (E. Noether (1922)) There exist forms $g_1,\dots,g_s$ in variables $A_{i_1\dots i_n}$ $(i_1 + \dots + i_n \le d)$ such that the above polynomial $f(X_1,\dots,X_n)$ is reducible over $\bar{K}$ or of degree $< d$ if and only if

$$g_j\{a_{i_1\dots i_n}\} = 0 \qquad (1 \le j \le s) .$$

Moreover, if $k = \binom{n+d-1}{n}$, then

(i) $\deg g_j \le k^{2^k}$ $\qquad (1 \le j \le s)$.

These forms depend only on n and d , and are independent of the field K in the sense that if $\operatorname{char} K = 0$, they are fixed forms with rational integer coefficients; while if $\operatorname{char} K = p\ (\neq 0)$, they are obtained by reducing the integral coefficients modulo p . In the case when $\operatorname{char} K = 0$,

(ii) $\|g_j\| \le 4^{k^{2^k}}$ $\qquad (1 \le j \le s)$.

Proof: We first dispose of the trivial cases. If $d = 1$, the forms may be taken to be just the variables corresponding to the coefficients of f. If $d \geq 2$ and $n = 1$, then f is always reducible over $\bar{K}$, so we may take $s = 1$ and g_1 identically zero. We may therefore assume that both $d \geq 2$ and $n \geq 2$, from which it follows that $k \geq 2$.

Observe that f is reducible or $\deg f < d$ if and only if $f = gh$ with $\deg g < d$, $\deg h < d$. Now suppose $f = gh$ where

$$g(X_1,\ldots,X_n) = \sum_{j_1+\ldots+j_n \leq d-1} b_{j_1\ldots j_n} X_1^{j_1} \ldots X_n^{j_n},$$

$$h(X_1\ldots,X_n) = \sum_{k_1+\ldots+k_n \leq d-1} c_{k_1\ldots k_n} X_1^{k_1} \ldots X_n^{k_n}.$$

Then the coefficients of f must have the form

$$a_{i_1\ldots i_n} = \sum_{j_1+k_1=i_1} \cdots \sum_{j_n+k_n=i_n} b_{j_1\ldots j_n} c_{k_1\ldots k_n}$$

for any $i_1,\ldots,i_n$ with $i_1+\ldots+i_n \leq 2d-2$ †). Let g be fixed, not identically zero, and consider the system of linear equations

$$(2.2) \quad c \cdot a_{i_1\ldots i_n} = \sum_{j_1+k_1=i_1} \cdots \sum_{j_n+k_n=i_n} b_{j_1\ldots j_n} c_{k_1\ldots k_n} \qquad (i_1+\ldots+i_n \leq 2d-2)$$

in c and the elements $c_{k_1\ldots k_n}$. If g divides f, then (2.2) has a solution with $c = 1$, hence has a non-trivial solution. Conversely, if (2.2) has a non-trivial solution, then if $c = 0$, we would obtain the contradictory result that $gh = 0$ while both $g \neq 0$ and $h \neq 0$. So in fact $c \neq 0$, and there is a solution of (2.2) with $c = 1$,

---

†) We set $a_{i_1\ldots i_n} = 0$ for $i_1 + \ldots + i_n > d$.

and hence g divides f.

We have shown that g divides f if and only if (2.2) has a non-trivial solution in the variables $c,\{c_{k_1\ldots k_n}\}$. The number of variables is $k+1$ with $k = \binom{n+d-1}{n}$. Therefore the condition that g divide f is that all the $(k+1) \times (k+1)$ determinants, say $\Delta_1,\ldots,\Delta_r$, of the system of linear equations (2.2) vanish. But each Δ_i is a form in the coefficients $b_{j_1\ldots j_n}$ of degree k, and the number of these coefficients is also k. We know from elimination theory, specifically Theorem 1A, that there exist forms $h_1,\ldots h_s$ in the coefficients of $\Delta_1,\ldots,\Delta_r$, such that the equations $\Delta_1 = \ldots = \Delta_r = 0$ have a non-trivial solution (in the $b_{j_1\ldots j_n}$'s) if and only if $h_1 = \ldots = h_s = 0$. Also by Theorem 1A,

$$\deg h_i = 2^{k-1} k^{2^{k-1}-1} \le k^{2^k} \qquad (1 \le i \le s). \tag{2.3}$$

If $\operatorname{char} K = 0$, it follows from Theorem 1D that

$$\|h_i\| \le 2^{2^{4k-4} k^{2^{k-1}}} \le 2^{k^{2^k}} \qquad (1 \le i \le s). \tag{2.4}$$

Now let g_i be obtained from h_i by substituting for the coefficients of the forms $\Delta_1,\ldots,\Delta_r$ their expressions in terms of the original coefficients $a_{i_1\ldots i_n}$ of f. Each such coefficient is linear in the $a_{i_1\ldots i_n}$ with norm at most $k!$ Combining (2.3), (2.4) with Lemmas 1B, 1E, we obtain

$$\deg g_i \leqq \deg h_i \leqq k^{2^k} \qquad (1 \leqq i \leqq s)$$

and

$$\|g_i\| \le \|h_i\| (k!)^{2^{k-1}} k^{2^{k-1}-1}$$

$$\le 2^{k^{2^k}} 2^{k^2} 2^{k-1} k^{2^{k-1}-1}$$

$$\le 2^{k^{2^k}} 2^{k^{2^{k-1}+k}}$$

$$\le 2^{k^{2^k}} 2^{k^{2^k}}$$

$$= 4^{k^{2^k}} .$$

COROLLARY 2B: (Ostrowski (1919)) Let $f(X_1,\dots,X_n)$ be a polynomial of degree $d > 0$ with rational integral coefficients. Suppose f is absolutely irreducible (i.e. irreducible over $\bar{\mathbb{Q}}$). Let p be a prime with

$$p > (4\|f\|)^{k^{2^k}} ,$$

where $k = \binom{n+d-1}{n}$. Then the reduced polynomial modulo p is again of degree d and absolutely irreducible (i.e. irreducible over $\bar{\mathbb{F}}_p$).

Proof: Let f be given by (2.1), where the coefficients $\{a_{i_1\dots i_n}\}$ are now integers. Since f is of degree d and absolutely irreducible, in the notation of Theorem 2A, not all the numbers $g_i(\{a_{i_1\dots i_n}\})$ are zero. Let us say $g_1(\{a_{i_1\dots i_n}\}) \ne 0$. We have the estimate

$$0 < |g_1(\{a_{i_1\dots i_n}\})| \le \|g_1\| . \|f\|^{k^{2^k}} \le (4\|f\|)^{k^{2^k}} .$$

So if $p > (4\|f\|)^{k^{2^k}}$, then the number $g_1(\{a_{i_1\dots i_n}\})$ is still non-zero modulo p . It follows, again by Theorem 2A , that the reduced polynomial modulo p is of degree d and absolutely irreducible.

COROLLARY 2C: Let $f(X,Y)$ be a polynomial with rational integer coefficients which is absolutely irreducible. If $N(p)$ denotes the number of solutions of the congruence

$$f(x,y) \equiv 0 \pmod{p} ,$$

then for large primes p ,

$$N(p) = p + O(p^{1/2}) .$$

Proof: Combine Corollary 2B with Theorem 1A of Chapter III .

§3. The absolute irreducibility of polynomials (II) .

Let K and L be two fields with $K \subseteq L$. The algebraic closure of K in L , denoted by K^o , is defined as the set of elements of L which are algebraic over K . Clearly K^o is a field and $K \subseteq K^o \subseteq L$.

THEOREM 3 A: Suppose $f(X_1,\ldots,X_m,Y)$ is a polynomial with coefficients in a field K , irreducible over K , and of degree $d > 0$ in Y . Further suppose that f is not a polynomial in only $X_1^p,\ldots,X_m^p,Y^p$ if K has characteristic $p \neq 0$. Let $\mathfrak{Y}$ be a quantity satisfying $f(X_1,\ldots,X_m,\mathfrak{Y}) = 0$, and let $L = K(X_1,\ldots,X_m,\mathfrak{Y})$. Let K^o be the algebraic closure of K in L . Then $[K^o : K]$ is a divisor of d and K^o is separable over K . Moreover, the polynomial $f(X_1,\ldots,X_m,Y)$ is absolutely irreducible if and only if $K^o = K$.

Theorems of this type are well known to algebraic geometers. See, e.g., Zariski (1944) . See also Corollary 6C in Ch. VI.

Example: Consider the polynomial $f(X,Y) = 2X^2 - Y^4$ over the field $K = \mathbb{Q}$ of rational numbers. Clearly $f(X,Y)$ is irreducible over $\mathbb{Q}$. Choose $\mathfrak{Y}$ so that $\mathfrak{Y}^4 = 2X^2$ and let $L = \mathbb{Q}(X,\mathfrak{Y})$. If we put $\alpha = \mathfrak{Y}^2/X$, then $\alpha^2 = 2$, so $\sqrt{2} \in \mathbb{Q}^0$. This means that $\mathbb{Q}$ is not algebraically closed in L , or $\mathbb{Q}^0 \neq \mathbb{Q}$. By Theorem 3 A , $f(X,Y)$ is not absolutely irreducible; in fact, we see directly that

$$f(X,Y) = (\sqrt{2}\,X - Y^2)(\sqrt{2}\,X + Y^2)$$

is a factorization of $f(X,Y)$ over $\mathbb{Q}(\sqrt{2})$.

Proof of Theorem 3 A. We begin with the following remark: If K^0 is algebraic over K of degree d , then $K^0(X_1,\ldots,X_m)$ is algebraic over $K(X_1,\ldots,X_m)$ of degree d , and vice versa. If K^0 is separable (or inseparable) over K , then $K^0(X_1,\ldots,X_m)$ is separable (or inseparable) over $K(X_1,\ldots,X_m)$, and conversely. This follows from the argument used in Lemma 2A of Chapter III .

Now observe that

$$(3.1) \quad K(X_1,\ldots,X_m) \subseteq K^0(X_1,\ldots,X_m) \subseteq K^0(X_1,\ldots,X_m,\mathfrak{Y}) = K(X_1,\ldots,X_m,\mathfrak{Y}) .$$

Since $K(X_1,\ldots,X_m,\mathfrak{Y})$ is an extension of $K(X_1,\ldots,X_m)$ of degree d , it follows that $[K^0(X_1,\ldots,X_m) : K(X_1,\ldots,X_m)]$ divides d , whence $[K^0 : K]$ divides d by the above remark.

If f is absolutely irreducible, then f is irreducible over K^0 . Hence $\mathfrak{Y}$ is algebraic of degree d over $K^0(X_1,\ldots,X_m)$; that is ,

$$[K^0(X_1,\ldots,X_m,\mathfrak{Y}) : K^0(X_1,\ldots,X_m)] = d .$$

From (3.1) it follows that $K(X_1,\ldots,X_m) = K^o(X_1,\ldots,X_m)$, so that $K = K^o$.

For the remainder of the proof, we shall tacitly assume that char $K = p \neq 0$. Actually the case when char $K = 0$ is simpler, and several steps may be omitted.

Let $f_1(X_1,\ldots,X_m,Y)$ be an irreducible factor of $f(X_1,\ldots,X_m,Y)$ over $\bar{K}$ such that

(3.2) $$f_1(X_1,\ldots,X_m,\mathfrak{Y}) = 0 .$$

We normalize f_1 by requiring that the leading coefficient (in some lexicographic ordering of the monomials) is 1 . Then every power of f_1 also has this property. Let K_1 be the field obtained from K by adjoining the coefficients of f_1 . Let a be the smallest positive integer such that every coefficient of f_1^a is separable over K . If b is a positive integer such that f_1^b has coefficients which are separable over K , then $a | b$: For if $b = at + r$ with $0 \leqq r < a$, then f_1^r has separable coefficients, and by the minimal choice of a , we have $r = 0$. Now $f_1^{p^\ell}$ has separable coefficients for some ℓ , hence $a | p^\ell$, and a itself must be a power of p . We have

$$K \subseteq K_1^s \subseteq K_1 ,$$

where K_1^s is the separable extension of K obtained from K by adjoining the coefficients of f_1^a .

The polynomial $g = f_1^a$ has coefficients in K_1^s and is irreducible over K_1^s , since its proper divisors (which would necessarily be powers of f_1) have coefficients which are not all separable over K , hence do not all lie in K_1^s. Now $g = f_1^a$ divides f^a, and since g is irreducible, g divides f. Write $\delta = [K_1^s : K]$ and let $g^{(1)}, g^{(2)}, \ldots, g^{(\delta)}$ be the distinct conjugates of g. Each $g^{(i)}$ divides f, so the product

$$g^{(1)} g^{(2)} \ldots g^{(\delta)} \mid f .$$

But this product has coefficients which are separable over K, and which are invariant under conjugation. Hence this product has coefficients in K. Since f is irreducible over K, there exists a constant $c \in K$ such that

$$f = c\, g^{(1)} g^{(2)} \ldots g^{(\delta)} .$$

If a were a positive power of p, then g would be a polynomial in $X_1^p, \ldots, X_m^p, Y^p$, hence each conjugate would be such a polynomial, and therefore f would be a polynomial in $X_1^p, \ldots, X_m^p, Y^p$. But this is impossible by hypothesis. Hence $a = 1$. It follows immediately that $K_1^s = K_1$, whence that K_1 is a separable extension of K.

Now $f = c\, f_1^{(1)} \ldots f_1^{(\delta)}$ has degree d in Y, so each factor $f_1^{(i)}$ has degree d/δ in Y. Hence by (3.2), $\mathfrak{Y}$ has degree d/δ over $K_1(X_1, \ldots, X_m)$. Since $[K_1 : K] = \delta$, it follows that $[K_1(X_1, \ldots, X_m, \mathfrak{Y}) : K(X_1, \ldots, X_m)] = d$. Since $K \subseteq K_1$, and since also $[K(X_1, \ldots, X_m, \mathfrak{Y}) : K(X_1, \ldots, X_m)] = d$, we have

$$K_1(X_1, \ldots, X_m, \mathfrak{Y}) = K(X_1, \ldots, X_m, \mathfrak{Y}) = L .$$

Thus K_1 is contained in L and is algebraic over K, whence $K_1 \subseteq K^o$.

Now f_1 was irreducible over K_1, in fact absolutely irreducible. By the part of the theorem already proved, $(K_1)^o = K_1$. But $(K_1)^o = K^o$, so $K_1 = K^o$, and K^o is separable over K. Finally, if $K^o = K$, then $K_1 = K$ and f is absolutely irreducible. This completes the proof.

We are now able to finish the

Proof of Lemma 2B of Chapter III: In the notation of that lemma, we need to show that if

$$[K(X,Z,\mathfrak{Y},\mathfrak{U}) : (K(X,Z)] = d^2 ,$$

then $f(X,Y)$ is absolutely irreducible. Suppose $f(X,Y)$ is not absolutely irreducible. By Theorem 3A, $K^o \neq K$. Let $[K^o(X) : K(X)] = u > 1$ and let $[K(X,\mathfrak{Y}) : K^o(X)] = v$, so that $uv = d$. In the chain $K(X,Z) \subsetneqq K^o(X,Z) \subsetneqq K^o(X,Z,\mathfrak{Y}) \subsetneqq K^o(X,Z,\mathfrak{Y},\mathfrak{U}) = K(X,Z,\mathfrak{Y},\mathfrak{U})$, the field extensions are of respective degrees u,v,v, so that

$$[K(X,Z,\mathfrak{Y},\mathfrak{U}) : K(X,Z)] = uv^2 < (uv)^2 = d^2 ,$$

which completes the proof.

In §2 of Chapter IV we introduced an equivalence relation for quadratic forms. We make a slight adjustment of that definition to define an equivalence for polynomials in n variables over a field K. We say that $f(\underline{\underline{X}}) \sim g(\underline{\underline{X}})$ if there is a non-singular $(n \times n)$ matrix T and a vector $\underline{\underline{t}}$, both having components in K, such that

$$f(\underline{\underline{X}}) = g(T\underline{\underline{X}} + \underline{\underline{t}}) .$$

This is clearly an equivalence relation.

<u>LEMMA 3B: Suppose $f(\underline{X}) \sim g(\underline{X})$. If f is irreducible over K (or absolutely irreducible), then so is g. Moreover, the total degrees of $f(\underline{X})$ and $g(\underline{X})$ are equal.</u>

Proof: Exercise. Notice that the first part of the lemma is a generalization of Lemma 2B of Chapter I.

Let $f(X_1,\ldots,X_n)$ be a polynomial over K. For $1 \le \ell \le n$, we will write

$$f(\widehat{X_1,\ldots,X_\ell}, X_{\ell+1},\ldots,X_n)$$

when the polynomial is to be interpreted as a polynomial in the variables $X_{\ell+1},\ldots,X_n$, with coefficients in the field $K(X_1,\ldots,X_\ell)$.

<u>LEMMA 9C: If $f(X_1,\ldots,X_n)$ is irreducible (over K), then $f(\widehat{X_1,\ldots,X_\ell}, X_{\ell+1},\ldots,X_n)$ is irreducible (over $K(X_1,\ldots,X_\ell)$).</u>

Proof: This follows from the unique factorization in $K[X_1,\ldots,X_\ell]$. The details are left as an exercise.

We remark that if $f(X_1,\ldots,X_n)$ is absolutely irreducible (i.e. irreducible over $\bar{K}$), it does not follow that $f(\widehat{X_1,\ldots,X_\ell}, X_{\ell+1},\ldots,X_n)$ is absolutely irreducible (i.e. irreducible over $\overline{K(X_1,\ldots,X_\ell)}$). In fact, if $\ell = n-1$, the new polynomial is a polynomial in one variable, which cannot be absolutely irreducible unless its degree is one. As another example, the polynomial

$$f(X_1,X_2,X_3) = X_2^2 - X_1X_3^2$$

is absolutely irreducible, while $f(\widehat{X_1},X_2,X_3)$ has the factorization

$$f(\widehat{X_1},X_2,X_3) = (X_2 - \sqrt{X_1}\,X_3)(X_2 + \sqrt{X_1}\,X_3)$$

over $\overline{K(X_1)}$.

THEOREM 3D: *Suppose* $f(X_1,\dots,X_n)$ *is a polynomial over an infinite field* K . *Suppose* f *is absolutely irreducible and of degree* $d > 0$. *Let* $1 \le \ell \le n-2$. *Then there is a polynomial* $g \sim f$ *such that*

$$g(\widehat{X_1,\dots,X_\ell}\,,X_{\ell+1},\dots,X_n)$$

is absolutely irreducible and of degree d (*in* $X_{\ell+1},\dots,X_n$) .

We shall need

LEMMA 3E: *Let* $J \subseteq L$ *be fields such that* L *is a finite separable algebraic extension of* J . *Then there are only finitely many fields* J' *with*

$$J \subseteq J' \subseteq L .$$

Proof: Let N be a finite separable algebraic normal extension of J with $L \subseteq N$. Let G be the Galois group of N over J , and let H be the Galois group of N over L . Then $H \subseteq G$. From Galois theory, we know that there is a one-one correspondence between fields J' with $J \subseteq J' \subseteq L$ and groups H' with $H \subseteq H' \subseteq G$. The number of such groups H' is finite, so the number of fields J' is finite.

Remark: Separability is essential in Lemma 3E. For let F be an algebraically closed (hence infinite) field of characteristic p . Take

$$J = F(X,Y) \subseteq L = J(X^{1/p}, Y^{1/p}) ,$$

and if $c \in F$, let

$$J_c' = J((X + cY)^{1/p}) = J(X^{1/p} + c^{1/p}Y^{1/p}) .$$

Clearly $J \subseteq J_c' \subseteq L$, but for different choices of $c \in F$ we get different fields J_c' , so that the collection of intermediate fields is infinite.

We begin the

<u>Proof of Theorem 3D</u>: We shall tacitly assume that $\operatorname{char} K = p \neq 0$, the proof for the case $\operatorname{char} K = 0$ being easier. First observe that $f(X_1,\ldots,X_n)$ is not a polynomial in $X_1^p,\ldots,X_n^p$, for if it were then

$$f(X_1,\ldots,X_n) = \sum_{i_1,\ldots,i_n} a_{i_1\ldots i_n} X_1^{pi_1} \cdots X_n^{pi_n}$$

$$= \left(\sum_{i_1,\ldots,i_n} a_{i_1\ldots i_n}^{1/p} X_1^{i_1} \cdots X_n^{i_n} \right)^p ,$$

contradicting the assumption that $f(X_1,\ldots,X_n)$ is absolutely irreducible. We change notation and write

$$f = f(X_1,\ldots,X_m,Y)$$

where $m = n - 1$. After a linear transformation of variables $(X_i' = X_i + c_iY ; i = 1,2,\ldots,m)$ we may suppose that f is of degree d in Y and separable in Y . Let $\mathfrak{Y}$ be a quantity satisfying

$$f(X_1,\ldots,X_m,\mathfrak{Y}) = 0 ,$$

and let $L = K(X_1,\ldots,X_m,\mathfrak{Y})$. For $c \in K$, put $X_1^{(c)} = X_1 + cX_m$. Construct the fields $K(X_1^{(c)})$ and $\left(K(X_1^{(c)})\right)^o$, the latter being the algebraic closure of $K(X_1^{(c)})$ in L .

LEMMA 3 F: For some $c \in K$,

$$\left(K(X_1^{(c)})\right)^o = K(X_1^{(c)}) .$$

Proof: For every $c \in K$ we have

$$K(X_1,\ldots,X_m) \subseteq \left(K(X_1^{(c)})\right)^o (X_2,\ldots,X_m) \subseteq L .$$

Note that L is a separable extension of $K(X_1,\ldots,X_m)$ of degree d . By Lemma 3E , there are only finitely many subfields of L containing $K(X_1,\ldots,X_m)$. Hence there exist two distinct elements $c,c' \in K$ such that

$$\left(K(X_1^{(c)})\right)^o (X_2,\ldots,X_m) = \left(K(X_1^{(c')})\right)^o (X_2,\ldots,X_m) ,$$

or

$$\left(K(X_1^{(c)})\right)^o (X_m)(X_2,\ldots,X_{m-1}) = \left(K(X_1^{(c')})\right)^o (X_m)(X_2,\ldots,X_{m-1}) .$$

But since $X_2,\ldots,X_{m-1}$ are algebraically independent over $K(X_1,X_m)$, it follows that

$$\left(K(X_1^{(c)})\right)^o (X_m) = \left(K(X_1^{(c')})\right)^o (X_m) .$$

For brevity we shall write $X = X_1^{(c)}$ and $Z = X_1^{(c')}$. By Theorem 3 A ,

$K(X_1^{(c)})^o$ is a finite separable extension of $K(X_1^{(c)})$, and hence there exists an element $\mathfrak{X}$ such that

$$\left(K(X_1^{(c)})\right)^o = \left(K(X)\right)^o = K(X,\mathfrak{X}) .$$

Similarly, there is a $\mathfrak{Z}$ with

$$\left(K(X_1^{(c')})\right)^o = \left(K(Z)\right)^o = K(Z,\mathfrak{Z}) .$$

Let $\mathfrak{X}$ have the defining equation $h_1(X,\mathfrak{X}) = 0$, where h_1 is irreducible over K; let $\mathfrak{Z}$ have the defining equation $h_2(Z,\mathfrak{Z}) = 0$, where h_2 is irreducible over K. Now by Theorem 3A and the absolute irreducibility of f, $K = K^o$, so that K is algebraically closed in L. It follows that K is algebraically closed in $K(X,\mathfrak{X})$ and in $K(Z,\mathfrak{Z})$. Then by Theorem 3A again, h_1 and h_2 are absolutely irreducible. Hence if $\mathfrak{X}$ is of degree d_1 over $K(X)$ and if $\mathfrak{Z}$ is of degree d_2 over $K(Z)$, then

$$[K(X,Z,\mathfrak{X},\mathfrak{Z}) : K(X,Z)] = d_1 d_2$$

by Lemma 2A of Chapter III. But we have

$$K(X,Z,\mathfrak{X}) = \left(K(X_1^{(c)})\right)^o (X_m) = \left(K(X_1^{(c')})\right)^o (X_m) = K(X,Z,\mathfrak{Z}) ,$$

so that

$$K(X,Z,\mathfrak{X}) = K(X,Z,\mathfrak{Z}) = K(X,Z,\mathfrak{X},\mathfrak{Z}) .$$

These three fields are extension of $K(X,Z)$ of respective degrees d_1, d_2 and $d_1 d_2$, so that $d_1 = d_2 = d_1 d_2$, and therefore $d_1 = d_2 = 1$. Hence $\left(K(X_1^{(c)})\right)^o = K(X_1^{(c)})$ and $\left(K(X_1^{(c')})\right)^o = K(X_1^{(c')})$, which proves the lemma.

We now conclude the proof of Theorem 3D. We may write

$$f(X_1,\ldots,X_m Y) = g(X_1^{(c)}, X_2,\ldots,X_m,Y)$$

where $c \in K$ is obtained from Lemma 3F and where

$$g(X,X_2,\ldots,X_m,Y) = f(X - cX_m, X_2,\ldots,X_m,Y) .$$

Clearly $g(X_1^{(c)},X_2,\ldots,X_m,\mathfrak{Y}) = 0$ and g is irreducible. But $g(\widehat{X_1^{(c)}},X_2,\ldots,X_m,Y)$ is absolutely irreducible (i.e., irreducible over $\overline{K(X_1^{(c)})}$) because $\left(K(X_1^{(c)})\right)^o = K(X_1^{(c)})$. By a change of notation, $g(\widehat{X_1},X_2,\ldots,X_m,Y)$ is absolutely irreducible. This new polynomial is clearly equivalent to f and is of degree d in Y. This process must now be repeated by setting $X_2^{(c)} = X_2 + cX_m$ with $c \in K$, etc., to obtain the result. Note that in the last step $X_\ell^{(c)} = X_\ell + cX_m$, hence that we certainly do need the condition $\ell \leqq m - 1 = n - 2$.

§ 4. The absolute irreducibility of polynomials (III).

Let K be a field. We have denoted by K^n the n-dimensional vector space over K consisting of n-tuples $(x_1,\ldots,x_n)$ with components in K. Suppose M is an m-dimensional linear manifold in K, where $1 \leq m \leq n$. Then M has a parameter representation

$$\underline{\underline{X}} = y_0 + U_1 \underline{\underline{y}}_1 + \ldots + U_m \underline{\underline{y}}_m ,$$

where $\underline{\underline{y}}_0, \underline{\underline{y}}_1, \ldots, \underline{\underline{y}}_m \in K^n$, with $\underline{\underline{y}}_1,\ldots \underline{\underline{y}}_m$ linearly independent, and where $U_1,\ldots,U_n$ are parameters. We write $\underline{\underline{X}} = L(\underline{\underline{U}})$. Suppose M has another parameter representation

$$\underline{\underline{X}} = L'(\underline{\underline{U}}') = \underline{\underline{y}}'_0 + U'_1\,\underline{\underline{y}}'_1 + \dots + U'_m\,\underline{\underline{y}}'_m .$$

Then $\underline{\underline{U}} = T\underline{\underline{U}}' + \underline{\underline{t}}$, where T is a non-singular $(m \times m)$-matrix over K and $\underline{\underline{t}} \in K^n$, hence $L(T\underline{\underline{U}}' + \underline{\underline{t}}) = L'(\underline{\underline{U}}')$. If $f(X_1,\dots,X_n)$ is a polynomial with coefficients in K and M is a linear manifold with parameter representation $L(\underline{\underline{U}})$, put

$$f_L(\underline{\underline{U}}) = f(L(\underline{\underline{U}})).$$

If L' is another parameter representation of M , then

$$f_{L'}(\underline{\underline{U}}') = f(L'(\underline{\underline{U}}')) = f(L(T\underline{\underline{U}}' + \underline{\underline{t}})) = f_L(T\underline{\underline{U}}' + \underline{\underline{t}}) .$$

Hence the polynomial f_L is determined by M up to equivalence in the sense of §3 . One can therefore speak of the "degree of f on M" and of the irreducibility or absolute irreducibility of f on M .

<u>LEMMA 4A:</u> <u>Suppose</u> $f(X_1,\dots,X_n)$ <u>has coefficients in an infinite field</u> K , <u>is of degree</u> $d > 0$ <u>and is absolutely irreducible. Let</u> $n \geq 3$ <u>and suppose that</u> m <u>is such that</u> $2 \leq m < n$. <u>Then there exists a linear manifold</u> M <u>of dimension</u> m <u>such that</u> f <u>is of degree</u> d <u>and absolutely irreducible on</u> M .

<u>Proof</u>: We may replace f by an equivalent polynomial. We may therefore assume by Theorem 3D that

$$f(\widehat{X_1,\dots,X_{n-m}},X_{n-m+1},\dots,X_n)$$

is of degree d (in $X_{n-m+1},\dots,X_n$) and is absolutely irreducible. By Theorem 2A , for polynomials in m variables of degree at most

d , there is a system of forms $g_1,\dots,g_s$ in the coefficients so that the polynomial is reducible or of degree $< d$ precisely if $g_1 = \dots = g_s = 0$. In our case, the coefficients are polynomials in $X_1,\dots,X_{n-m}$, so that we may write

$$g_i = g_i(X_1,\dots,X_{n-m}) \qquad (1 \le i \le s) .$$

Since $f(\widehat{X_1,\dots,X_{n-m}},X_{n-m+1},\dots,X_n)$ is of degree d and is absolutely irreducible, we must have some $g_i(X_1,\dots,X_{n-m}) \neq 0$, say for simplicity $g_1(X_1,\dots,X_{n-m}) \neq 0$. Since K is infinite there exist elements $t_1,\dots,t_{n-m} \in K$ such that $g_1(t_1,\dots,t_{n-m}) \neq 0$. Then the polynomial

$$f(t_1,\dots,t_{n-m},X_{n-m+1},\dots,X_n)$$

in variables $X_{n-m+1},\dots,X_n$ is of degree d and absolutely irreducible. This means simply that the polynomial f on the manifold M given by

$$x_1 = t_1,\dots,x_{n-m} = t_{n-m}$$

is of degree d and absolutely irreducible, which proves the lemma.

Let M be a linear manifold of dimension $m \geqq 2$ with parameter representation

$$(4.1) \qquad \underline{\underline{X}} = L(\underline{\underline{U}}) = \underline{\underline{y}}_0 + U_1\, \underline{\underline{y}}_1 + \dots + U_m\, \underline{\underline{y}}_m .$$

The polynomial f_L is absolutely irreducible and of degree d precisely if not all of certain froms $g_1,\dots,g_s$ in the coefficients of f_L vanish. We have $g_i = g_i(\underline{\underline{y}}_0,\dots,\underline{\underline{y}}_m)$, where $g_i(\underline{\underline{Y}}_0,\dots,\underline{\underline{Y}}_m)$ are polynomials in $n(m+1)$ variables. Since there exists a manifold M on which f is of degree d and absolutely irreducible, not all these polynomials $g_i(\underline{\underline{Y}}_0,\underline{\underline{Y}}_1,\ldots,\underline{\underline{Y}}_n)$ are identically zero.

Let F be a subfield of K. We shall say that a linear manifold M in K^n is generic if it has[†)] a parameter representation (4.1) where the $n(m+1)$ components of $\underline{\underline{y}}_0,\underline{\underline{y}}_1,\ldots,\underline{\underline{y}}_n$ are algebraically independent over F. (That is, they satisfy no non-trivial polynomial equation in $n(m+1)$ variables with coefficients in F). More precisely, one should say that M is generic over F. Suppose $f(X_1,\ldots,X_n)$ has coefficients in F and is absolutely irreducible. Then some $g_i(\underline{\underline{Y}}_0,\ldots,\underline{\underline{Y}}_n) \neq 0$, whence $g_i(\underline{\underline{y}}_0,\ldots,\underline{\underline{y}}_n) \neq 0$ if the components of $\underline{\underline{y}}_0,\underline{\underline{y}}_1,\ldots,\underline{\underline{y}}_n$ are algebraically independent over F. Thus f is absolutely irreducible on M. We thus have

THEOREM 4B: Let $f(\underline{\underline{X}}) \in F[\underline{\underline{X}}]$ be absolutely irreducible and of degree d. Then on a generic linear manifold M of dimension m $(2 \le m \le n)$, the restriction of f is again absolutely irreducible and of degree d.

This theorem, or rather a generalization of it, is sometimes called Bertini's Theorem. It is connected with work of the Italian geometer Bertini (1892).

Example: Take $n=3$ and $m=2$. The polynomial

$$f(X_1,X_2,X_3) = X_1^2 + X_2^2 - X_3^2 - 1$$

defines a hyperboloid of one shell in 3-space. The intersection of this hypersurface with a plane (a 2-dimensional linear manifold) can be an ellipse, a hyperbola, a parabola, or if the plane is tangent[*)]

*) this includes the case when the plane is "tangent to a point at infinity".

†) Note that the parameter representation is not unique.

to the surface, two lines. The restriction of f to a plane is reducible precisely if the intersection consists of two lines; that is, precisely if the plane is tangent to the surface. It can be shown that the tangent planes are the planes

$$a_1x_1 + a_2x_2 + a_3x_3 + a_0 = 0$$

with $a_1^2 + a_2^2 - a_3^2 - a_0^2 = 0$. The planes with $a_0 = 0$ are tangent to an infinite point of the hyperboloid, and the intersection of the hyperboloid with such a plane consists of two parallel lines (i.e., two lines which intersect at an infinite point). The other tangent planes have an intersection with the hyperboloid which consists of two intersecting lines (i.e., lines whose intersection is a finite point).

THEOREM 4C: Let $f(X_1,\ldots,X_n)$ be a polynomial over F_q of degree $d > 0$ which is absolutely irreducible. Let $n \ge 3$ and let A be the number of 2-dimensional linear manifolds $M^{(2)}$. Let B denote the number of manifolds $M^{(2)}$ on which f is not of degree d or is not absolutely irreducible. Let $\Psi = 2dk^{2^k}$ where $k = \binom{d+1}{2}$. Then

$$B/A \le \Psi/q .$$

Proof: Every linear manifold $M^{(2)}$ has a parameter representation

$$\underline{\underline{X}} = \underline{\underline{y}}_0 + U_1 \underline{\underline{y}}_1 + U_2 \underline{\underline{y}}_2 ,$$

where $\underline{\underline{y}}_0, \underline{\underline{y}}_1, \underline{\underline{y}}_2 \in F_q^n$, and $\underline{\underline{y}}_1$ and $\underline{\underline{y}}_2$ are linearly independent. If A' is the number of such parameter representations, then

$$A' = q^n(q^n - 1)(q^n - q) \ge \frac{1}{2} q^{3n} .$$

But each linear manifold $M^{(2)}$ has

$$D = q^2(q^2 - 1)(q^2 - q)$$

different parameter representations, whence $A = A'/D$. Now on a manifold $M^{(2)}$,

$$f_L(\underline{\underline{X}}) = f(\underline{\underline{y}}_0 + U_1 \underline{\underline{y}}_1 + U_2 \underline{\underline{y}}_2)$$

is a polynomial in U_1, U_2 . By Theorem 2A , there are forms $g_1,\ldots,g_s$ in the coefficients of this polynomial such that $g_1 = \ldots g_s = 0$ is equivalent to the polynomial being of degree $< d$ or irreducible. The degree of each g_i was at most

$$k^{2^k} = \Psi' ,$$

say, where $k = \binom{d+1}{2}$. (Note that f_L is a polynomial in 2 variables). The coefficients of $f(\underline{\underline{y}}_0 + U_1 \underline{\underline{y}}_1 + U_2 \underline{\underline{y}}_2)$ are polynomials in the coordinates of $\underline{\underline{y}}_0, \underline{\underline{y}}_1, \underline{\underline{y}}_2$ of degree at most d . Substituting these coefficients into $g_1,\ldots,g_s$, we obtain polynomials $h_1,\ldots h_s$ in the coordinates of $\underline{\underline{y}}_0, \underline{\underline{y}}_1, \underline{\underline{y}}_2$, each of degree at most $d\Psi'$, and having the property that $f(\underline{\underline{y}}_0 + U_1 \underline{\underline{y}}_1 + U_2 \underline{\underline{y}}_2)$ is of degree $< d$ or reducible if and only if $h_i(\underline{\underline{y}}_0, \underline{\underline{y}}_1, \underline{\underline{y}}_2) = 0$ for $i = 1,\ldots,s$. Since the restriction of f to a generic manifold $M^{(2)}$ is absolutely irreducible, some $h_i = h_i(\underline{\underline{Y}}_0, \underline{\underline{Y}}_1, \underline{\underline{Y}}_2)$, say h_1, is not identically zero. By Lemma 3A of Chapter IV, the number of $\underline{\underline{y}}_0, \underline{\underline{y}}_1, \underline{\underline{y}}_2$ with $h_1(\underline{\underline{y}}_0, \underline{\underline{y}}_1, \underline{\underline{y}}_2) = 0$ is at most $d\,\Psi' q^{3n-1}$. But since each $M^{(2)}$ has D representations,

$$B \le d\,\Psi'\, q^{3n-1}/D .$$

Hence

$$B/A \le d\,\Psi'\, q^{3n-1}/A' \le 2d\,\Psi'/q = \Psi/q .$$

§ 5. The number of zeros of absolutely irreducible polynomials in n variables.

In this section we shall allow the symbols $\omega(q,d)$ and $\chi(d)$ to take on either one of the following interpretations:

(i) $\omega(q,d) = \sqrt{2}\, d^{5/2} q^{1/2}$, $\chi(d) = 250\, d^5$,

(ii) $\omega(q,d) = (d-1)(d-2)q^{1/2} + d^2$, $\chi(d) = 1$.

So if $f(X,Y)$ is a polynomial with coefficients in F_q , absolutely irreducible and of degree $d > 0$, then

$$|N - q| < \omega(q,d) \tag{5.1}$$

whenever $q > \chi(d)$, where N is the number of zeros of $f(X,Y)$. With interpretation (i) , this statement has been proved as Theorem 1A of Chapter III. However the statement also holds under interpretation (ii), as follows from the study of the zeta function of the curve $f(x,y)$ (Weil (1948a), Bombieri (1973)), and as may be known to a more sophisticated reader.

THEOREM 5A; *Suppose* $f(X_1,\ldots,X_n)$ *is a polynomial over* F_q *of total degree* $d > 0$ *and absolutely irreducible. Let* N *be the number of zeros of* f *in* F_q^n . *Then*

$$|N - q^{n-1}| \le q^{n-2}(\omega(q,d) + 2d\Psi) , \tag{5.2}$$

where Ψ *was defined in Theorem* 4C.

If interpretation (i) is used, we obtain

$$|N - q^{n-1}| \le q^{n-2}(\sqrt{2}\, d^{5/2} q^{1/2} + 2d\Psi) .$$

If we use interpretation (ii) , then

$$|N - q^{n-1}| \le q^{n-2}\left((d-1)(d-2)q^{1/2} + d^2 + 2d\,\Psi\right)$$

$$\le (d-1)(d-2)q^{n-(3/2)} + 3d\,\Psi\, q^{n-2} .$$

This theorem is due to Lang and Weil (1954) , and also Nisnevich (1954) . However, no value of the constant $2d\,\Psi$ was given . We now begin the

Proof: For a 2-dimensional linear manifold $M^{(2)}$ in F_q^n , let $N(M^{(2)})$ be the number of zeros of f on $M^{(2)}$. Every point of F_q^n lies on exactly

$$E = \frac{(q^n - 1)(q^n - q)}{(q^2 - 1)(q^2 - q)}$$

manifolds $M^{(2)}$. Thus

(5.3) $$N = \frac{1}{E} \sum_{M^{(2)}} N(M^{(2)}) .$$

Observe that by the property of $\omega(q,d)$ discussed above and by Lemma 3A of Chapter IV, we have for $q > \chi(d)$,

(5.4) $$|N(M^{(2)}) - q| \le \begin{cases} \omega(q,d) & \text{if } f \text{ is absol. irred. on } M^{(2)}, \\ dq & \text{if } f \text{ is not identically zero on } M^{(2)}, \\ q^2 & \text{if } f = 0 \text{ identically on } M^{(2)} . \end{cases}$$

LEMMA 5B: Let $f(X_1,\ldots,X_n)$ be a polynomial over F_q , of degree $d > 0$ and irreducible. Suppose f is not equivalent to a polynomial $g(X_1,\ldots,X_{n-2})$, where only $n - 2$ variables appear. As in Theorem 4C , let A be the number of 2-dimensional linear manifolds $M^{(2)}$. Let C

be the number of manifolds $M^{(2)}$ where f is identically zero. Then

$$C/A \le d^3/q^2 .$$

Proof: Consider the planes $M^{(2)}$ parallel to the plane $x_1 = \dots = x_{n-2} = 0$; these number $A^* = q^{n-2}$. Let C^* be the number of those parallel planes on which f is identically zero. A typical plane of this type is

$$M^{(2)} : x_1 = c_1 , \dots , x_{n-2} = c_{n-2} .$$

The polynomial f can, of course, be written as

$$f(X_1,\dots,X_n) = \sum_{i,j} p_{ij}(X_1,\dots,X_{n-2}) X_{n-1}^i X_n^j .$$

If f is identically zero on $M^{(2)}$, then

$$p_{ij}(c_1,\dots,c_{n-2}) = 0$$

for all i and j . If these polynomials p_{ij} have a common factor $g(X_1,\dots,X_{n-2})$ of positive degree, then g divides f and, since f is irreducible , $f = cg$. But by hypothesis f is not a polynomial in only $n - 2$ variables, hence the p_{ij} have no proper common factor. By Lemma 3D of Chapter IV, the number of common zeros $(c_1,\dots,c_{n-2})$ of the polynomials p_{ij} is at most $d^3 q^{n-4}$. It follows that $C^* \le d^3 q^{n-4}$ and

$$C^*/A^* \le d^3/q^2 .$$

The same argument holds for planes parallel to any given plane, and the result follows.

We now continue the

Proof of Theorem 5A: The proof is by induction on n. The case $n = 1$ is completely trivial, and the case $n = 2$ holds by what we said above. If $f \sim g$ where g is a polynomial in $n - 2$ variables, then the number of zeros of f is q^2 times the number N' of zeros of g in F_q^{n-2}. So by induction

$$|N' - q^{n-3}| \le q^{n-4}\left(\omega(q,d) + 2d\Psi\right),$$

whence (5.2). We may therefore suppose that f is not equivalent to a polynomial in $n - 2$ variables. Assume at first that $q > \chi(d)$. From (5.3) and (5.4) we find that

$$\left|N - \frac{1}{E}\sum_{M^{(2)}} q\right| \le \frac{1}{E}\left(\omega(q,d)\sum_{M^{(2)}} 1 + dq \sum_{\substack{M^{(2)} \\ f \text{ not absol. irred.}}} 1 + q^2 \sum_{\substack{M^{(2)} \\ f \equiv 0 \text{ on } M^{(2)}}} 1\right).$$

In our established notation, it follows that

$$\begin{aligned} |N - q^{n-1}| &\le \frac{1}{E}\left(\omega(q,d)A + dqB + q^2C\right) \\ &= (A/E)\left(\omega(q,d) + dq(B/A) + q^2(C/A)\right) \\ &\le q^{n-2}\left(\omega(q,d) + d\Psi + d^3\right) \\ &\le q^{n-2}\left(\omega(q,d) + 2d\Psi\right). \end{aligned}$$

On the other hand if $q < \chi(d)$, then $q^2 < 2d\Psi$, whence

$$|N - q^{n-1}| < q^n < q^{n-2}\left(\omega(q,d) + 2d\Psi\right).$$

COROLLARY 5C: Suppose $f(X_1,\dots,X_n)$ is a polynomial with rational integer coefficients which is of degree d and absolutely irreducible. For primes p, let $N(p)$ be the number of solutions of the congruence

$$f(x_1,\dots,x_n) \equiv 0 \pmod{p} .$$

Then as $p \to \infty$,

$$N(p) = p^{n-1} + O\left(p^{n-(3/2)}\right).$$

Proof: The proof is a combination of Theorem 5A and Corollary 2B .

The error terms of Theorem 5A in the two possible interpretations are

$$\sqrt{2}\ d^{5/2} q^{n-(3/2)} + O\left(q^{n-2}\right)$$

and

(5.5) $$(d-1)(d-2)q^{n-(3/2)} + O\left(q^{n-2}\right) .$$

It may be shown (Weil (1948a)) that when $n = 2$, the exponent $\frac{1}{2}$ in the error term $(d-1)(d-2)q^{\frac{1}{2}} + O(1)$ is best possible. Also the constant $(d-1)(d-2)$ is best possible. If $g(X,Y)$ is a polynomial in 2 variables with N' zeros , then the polynomial $f(X_1,\dots,X_n) = g(X_1,X_2)$ in n variables has $N = N'q^{n-2}$ zeros. Hence the exponent $n - (3/2)$ and the constant $(d-1)(d-2)$ in (5.5) are best possible for every n .

On the other hand the constant $2d\Psi$ in (5.2) is certainly too large. This is especially bad if one wants to estimate how large q must be in order that $N > 0$. With (5.2) one needs that q is certainly larger than $2d\Psi$, hence that q is very large as a function of d .

Schmidt (1973) applied the method of Stepanov directly to equations in n variables and obtained

$$N > q^{n-1} - 3d^3 q^{n-(3/2)} \quad \text{provided} \quad q > c_0 n^3 d^6 \quad ,$$

if (5.1) is used with $\omega(q,d)$ given by (i), and

$$N > q^{n-1} - (d-1)(d-2)q^{n-(3/2)} - 6d^2 q^{n-2} \quad \text{provided}$$
$$q > c_0(\varepsilon) n^3 d^{5+\varepsilon}$$

if (5.1) is used with $\omega(q,d)$ given by (ii) .

Much more is true for "non-singular" hypersurfaces by the deep work of Deligne (1973)[+)].

+) But see the remark in the Preface.

VI. Rudiments of Algebraic Geometry. The Number of Points in Varieties over Finite Fields.

General References: Artin (1955), Lang (1958), Shafarevich (1974), Mumford ( ) .

§1. Varieties.

THEOREM 1A. Let k be a field. Let $X_1,\dots,X_n$ be variables.

(i) In the ring $k[X_1,X_2,\dots,X_n]$, every ideal has a finite basis.

(ii) In this ring the ascending chain condition holds, i.e., if $\mathfrak{A}_1 \subseteq \mathfrak{A}_2 \subseteq \dots$ is an ascending sequence of ideals, then for some m, $\mathfrak{A}_m = \mathfrak{A}_{m+1} = \dots$.

(iii) Every non-empty set of ideals in this ring which is partially ordered by set inclusion, has at least one maximal element.

Statement (i) is the Hilbert Basis Theorem (Hilbert 1888). It is well known that the three conditions (i), (ii), (iii) for a ring R are equivalent. A ring satisfying these conditions is called Noetherian. A proof of this Theorem may be found in books on algebra, e.g. Van der Waerden (1955), Kap. 12 or Zariski-Samuel (1958), Ch. IV, and will not be given here.

If k, K are fields such that $k \subseteq K$, the transcendence degree of K over k, written tr. deg. K/k, is the maximum number of elements in K which are algebraically independent over k.

In what follows, k, Ω will be fields such that $k \subseteq \Omega$, the tr. deg $\Omega/k = \infty$, and Ω is algebraically closed. We call k the ground field, and Ω the universal domain. For example, we may take

$k = \mathbb{Q}$ (the rationals), $\Omega = \mathbb{C}$ (the complex numbers). Or $k = F_q$, the finite field of a q elements, $\Omega = \overline{F_q(X_1, X_2, \ldots)}$, i.e. the algebraic closure of $F_q(X_1, X_2, \ldots)$.

Consider Ω^n, the space of n-tuples of elements in Ω. Suppose $\mathfrak{J}$ is an ideal in $k[X_1, \ldots, X_n] = k[\underline{X}]$. Let $A(\mathfrak{J})$ be the set of $\underline{x} = (x_1, \ldots, x_n) \in \Omega^n$ having $f(\underline{x}) = 0$ for every $f(\underline{X}) \in \mathfrak{J}$. Every set $A(\mathfrak{J})$ so obtained is called an *algebraic set*. More precisely, it is a k-algebraic set. If we have such an ideal $\mathfrak{J}$, then by Theorem 1A, there exists a basis of $\mathfrak{J}$ consisting of a finite number of polynomials, say $f_1(\underline{X}), \ldots, f_m(\underline{X})$. Therefore $A(\mathfrak{J})$ can also be characterized as the set of $\underline{x} \in \Omega^n$ with $f_1(\underline{x}) = \ldots = f_m(\underline{x}) = 0$. Note that if $\mathfrak{J}_1 \subseteq \mathfrak{J}_2$, then $A(\mathfrak{J}_1) \supseteq A(\mathfrak{J}_2)$.

Examples: (1) Let $k = \mathbb{Q}$, $\Omega = \mathbb{C}$, $n = 2$, and $\mathfrak{J}$ the ideal generated by $f(X_1, X_2) = X_1^2 + X_2^2 - 1$. Then $A(\mathfrak{J})$ is the unit circle.

(2) Again let $k = \mathbb{Q}$, $\Omega = \mathbb{C}$, $n = 2$, and take $\mathfrak{J}$ to be the ideal generated by $f(X_1, X_2) = X_1^2 - X_2^2$. Then $A(\mathfrak{J})$ consists of the two intersecting lines $x_2 = x_1$, $x_2 = -x_1$.

THEOREM 1B. (i) *The empty set $\emptyset$ and Ω^n are algebraic sets.*

(ii) *A finite union of algebraic sets is an algebraic set.*

(iii) *An intersection of an arbitrary number of algebraic sets is an algebraic set.*

Proof: (i) If $\mathfrak{J} = k[X_1, \ldots, X_n]$, then $A(\mathfrak{J}) = \emptyset$. If $\mathfrak{J} = (0)$, i.e. the principal ideal generated by the zero polynomial, then $A(\mathfrak{J}) = \Omega^n$.

(ii) It is sufficient to show that the union of two algebraic sets is again an algebraic set. Suppose A is the algebraic set given by the equations $f_1(\underline{x}) = \dots = f_\ell(\underline{x}) = 0$, B is the algebraic set given by the equations $g_1(\underline{x}) = \dots = g_m(\underline{x}) = 0$. Then $A \cup B$ is the set of $\underline{x} \in \Omega^n$ with $f_1(\underline{x})\, g_1(\underline{x}) = f_1(\underline{x})\, g_2(\underline{x}) = \dots = f_\ell(\underline{x})\, g_m(\underline{x}) = 0$.

(iii) Let A_α , $\alpha \in I$, where I is any indexing set, be a collection of algebraic sets. Suppose that $A_\alpha = A(\mathfrak{J}_\alpha)$, where $\mathfrak{J}_\alpha$ is an ideal in $k[\underline{X}]$. We claim that

(1.1)
$$\bigcap_{\alpha \in I} A(\mathfrak{J}_\alpha) = A\left(\sum_{\alpha \in I} \mathfrak{J}_\alpha\right),$$

where $\sum_{\alpha \in I} \mathfrak{J}_\alpha$ is the ideal consisting of sums $f_1(\underline{X}) + \dots + f_\ell(\underline{X})$ with each $f_i(\underline{X})$ in $\mathfrak{J}_\alpha$ for some $\alpha \in I$. To prove (1.1) , suppose that $\underline{x} \in \bigcap_{\alpha \in I} A(\mathfrak{J}_\alpha)$. Then for each $\alpha \in I$, $\underline{x} \in A(\mathfrak{J}_\alpha)$, whence $f(\underline{x}) = 0$ if $f \in \mathfrak{J}_\alpha$. Therefore $f(\underline{x}) = 0$ if $f \in \sum_{\alpha \in I} \mathfrak{J}_\alpha$. Hence $\underline{x} \in A\left(\sum_{\alpha \in I} \mathfrak{J}_\alpha\right)$. Conversely, if $\underline{x} \in A\left(\sum_{\alpha \in I} \mathfrak{J}_\alpha\right)$, then $f(\underline{x}) = 0$ if $f \in \sum_{\alpha \in I} \mathfrak{J}_\alpha$. So for any $\alpha \in I$, if $f \in \mathfrak{J}_\alpha$, then $f(\underline{x}) = 0$. Thus, $\underline{x} \in A(\mathfrak{J}_\alpha)$ for all α , or $\underline{x} \in \bigcap_{\alpha \in I} A(\mathfrak{J}_\alpha)$. This proves (1.1) . It follows that $\bigcap A_\alpha = \bigcap A(\mathfrak{J}_\alpha)$ is an algebraic set.

In Ω^n we can now introduce a topology by defining the closed sets as the algebraic sets. This topology is called the Zariski Topology. As usual, the closure of a set M is the intersection of the closed sets containing M . It is the smallest closed set containing M and is denoted by $\overline{M}$.

Let M be a subset of Ω^n . We write $\mathfrak{J}(M)$ for the ideal of all polynomials $f(\underline{X})$ which vanish on M, i.e., all polynomials $f(\underline{X})$

such that $f(\underline{\underline{x}}) = 0$ for every $\underline{\underline{x}} \in M$. It is clear that if $M_1 \subseteq M_2$, then $\mathfrak{I}(M_1) \supseteq \mathfrak{I}(M_2)$.

THEOREM 1C. $\overline{M} = A\mathfrak{I}(M))$.

Proof: Clearly $A(\mathfrak{I}(M))$ is a closed set containing M. Therefore it is sufficient to show that $A(\mathfrak{I}(M))$ is the smallest closed set containing M. Let T be a closed set containing M; say $T = A(\mathfrak{J})$. Since $T \supseteq M$, it follows that $\mathfrak{J} \subseteq \mathfrak{I}(T) \subseteq \mathfrak{I}(M)$, so that

$$T = A(\mathfrak{J}) \supseteq A\mathfrak{I}(M)) .$$

Remark: If S is an algebraic set, then it follows from Theorem 1C that $S = A\mathfrak{I}(S))$.

If $\mathfrak{A}$ is an ideal, define the radical of $\mathfrak{A}$, written $\sqrt{\mathfrak{A}}$, to consist of all $f(\underline{\underline{X}})$ such that for some positive integer m, $f^m(\underline{\underline{X}}) \in \mathfrak{A}$. The radical of $\mathfrak{A}$ is again an ideal. For if $f(\underline{\underline{X}}), g(\underline{\underline{X}}) \in \sqrt{\mathfrak{A}}$, then there exist positive integer m, ℓ such that $f^m(\underline{\underline{X}}), g^{\ell}(\underline{\underline{X}}) \in \mathfrak{A}$. Thus by the Binomial Theorem, $(f(\underline{\underline{X}}) \pm g(\underline{\underline{X}}))^{m+\ell} \in \mathfrak{A}$, so that $f(\underline{\underline{X}}) \pm g(\underline{\underline{X}}) \in \sqrt{\mathfrak{A}}$. Also, for any $h(\underline{\underline{X}})$ in $k[\underline{\underline{X}}]$, $(h(\underline{\underline{X}})\ f(\underline{\underline{X}}))^m \in \mathfrak{A}$, so that $h(\underline{\underline{X}}) f(\underline{\underline{X}}) \in \sqrt{\mathfrak{A}}$.

If $\mathfrak{P}$ is a prime ideal, then $\sqrt{\mathfrak{P}} = \mathfrak{P}$, since if $f(\underline{\underline{X}}) \in \sqrt{\mathfrak{P}}$, then $f^m(\underline{\underline{X}}) \in \mathfrak{P}$, which implies that $f(\underline{\underline{X}}) \in \mathfrak{P}$.

THEOREM 1D. Let $\mathfrak{A}$ be an ideal in $k[\underline{\underline{X}}]$. Then

$$\mathfrak{I}(A(\mathfrak{A})) = \sqrt{\mathfrak{A}} .$$

Example: Let $k = \mathbf{Q}$, $\Omega = \mathbf{C}$, $n = 2$, and $\mathfrak{A}$ the principal ideal generated by $f(X_1, X_2) = (X_1^2 + X_2^2 - 1)^3$. Then $A(\mathfrak{A})$ is the unit circle, and $\mathfrak{I}(A(\mathfrak{A})) = (X_1^2 + X_2^2 - 1)$. Thus $\sqrt{\mathfrak{A}} = (X_1^2 + X_2^2 - 1)$, the ideal generated by $X_1^2+X_2^2-1$.

Before proving Theorem 1D we need two lemmas.

LEMMA 1E. Given a prime ideal $\mathfrak{P} \neq k[\underline{\underline{X}}]$, there exists an $\underline{\underline{x}} \in \Omega^n$ with $\mathfrak{I}(\underline{\underline{x}}) = \mathfrak{P}$.

<u>Proof.</u> Form the natural homomorphism from $k[\underline{\underline{X}}]$ to the quotient ring $k[\underline{\underline{X}}]/\mathfrak{P}$. Since $\mathfrak{P} \cap k = \{0\}$, the natural homomorphism is an isomorphism on k. Thus we may consider $k[\underline{\underline{X}}]/\mathfrak{P}$ as an extension of k, and the natural homomorphism restricted to k becomes the identity map. Thus our homomorphism is a k-homomorphism. Let the image of X_i be $\xi_i (i = 1,\ldots,n)$. The natural homomorphism is then a homomorphism from $k[X_1,\ldots,X_n]$ onto $k[\xi_1,\ldots,\xi_n]$ with kernel $\mathfrak{P}$. Since $\mathfrak{P}$ was a prime ideal, $k[\xi_1,\ldots,\xi_n]$ is an integral domain.

Try to replace ξ_i by $x_i \in \Omega$. If, say, $\xi_1,\ldots,\xi_d$ are algebraically independent over k with $\xi_{d+1},\ldots,\xi_n$ algebraically dependent on them, choose $x_1,\ldots,x_d \in \Omega$ algebraically independent over k. Then $k(\xi_1,\ldots,\xi_d)$ is k-isomorphic to $k(x_1,\ldots,x_d)$. Also, ξ_{d+1} is algebraic over $k(\xi_1,\ldots,\xi_d)$, and so satisfies a certain irreducible equation with coefficients in $k(\xi_1,\ldots,\xi_d)$. Choose x_{d+1} in Ω such that it satisfies the corresponding equation as ξ_{d+1} but with coefficients in $k(x_1,\ldots,x_d)$. Then $k(\xi_1,\ldots,\xi_{d+1})$ is k-isomorphic to $k(x_1,\ldots,x_{d+1})$. There is a k-isomorphism with $\xi_i \to x_i$ $(i = 1,\ldots,d+1)$.

Continuing in this manner, we can find $x_1,\ldots,x_n \in \Omega$ such that $k(\xi_1,\ldots,\xi_n)$ is k-isomorphic to $k(x_1,\ldots,x_n)$. There is an isomorphism α with $\alpha(\xi_i) = x_i$ $(i = 1,\ldots,n)$.

Composing the natural homomorphism with the isomorphism α we obtain a homomorphism

$$\varphi\colon k[X_1,\ldots,X_n] \to k[x_1,\ldots,x_n]$$

with kernel $\mathfrak{P}$. Write $\underline{\underline{x}} = (x_1,\ldots,x_n)$.

Now $\mathfrak{J}(\underline{\underline{x}}) = \mathfrak{P}$, for $f(\underline{\underline{x}}) = 0$ precisely if $\varphi(f(\underline{\underline{X}})) = 0$, which is true if $f(\underline{\underline{X}}) \in \mathfrak{P}$.

<u>LEMMA 1F.</u> <u>Let</u> $\mathfrak{S}$ <u>be a non-empty subset of</u> $k[\underline{\underline{X}}]$ <u>which is closed under multiplication and doesn't contain zero.</u> <u>Let</u> $\mathfrak{P}$ <u>be an ideal</u>

which is maximal with respect to the property that $\mathfrak{P} \cap \mathfrak{C} = \emptyset$. _Then_ $\mathfrak{P}$ _is a prime ideal._

Proof: Suppose $f(\underline{\underline{X}})g(\underline{\underline{X}}) \in \mathfrak{P}$ but that $f(\underline{\underline{X}})$ and $g(\underline{\underline{X}})$ are not in $\mathfrak{P}$. Let $\mathfrak{A} = (\mathfrak{P}, f(\underline{\underline{X}}))^{*}$, so that $\mathfrak{A}$ properly contains $\mathfrak{P}$. Since $\mathfrak{P}$ is maximal with respect to the property that $\mathfrak{P} \cap \mathfrak{C} = \emptyset$, it follows that $\mathfrak{A} \cap \mathfrak{C} \neq \emptyset$. So there exists a $c(\underline{\underline{X}}) = p(\underline{\underline{X}}) + h(\underline{\underline{X}})\, f(\underline{\underline{X}})$, where $c(\underline{\underline{X}}) \in \mathfrak{C}$, $p(\underline{\underline{X}}) \in \mathfrak{P}$, $h(\underline{\underline{X}}) \in k[\underline{\underline{X}}]$. Similarly, there exists a $c'(\underline{\underline{X}}) = p'(\underline{\underline{X}}) + h'(\underline{\underline{X}})\, g(\underline{\underline{X}})$, where $c'(X) \in \mathfrak{C}$, $p'(\underline{\underline{X}}) \in \mathfrak{P}$, $h'(\underline{\underline{X}}) \in k[\underline{\underline{X}}]$. Then

$$c'(\underline{\underline{X}})\, c(\underline{\underline{X}}) = (p'(\underline{\underline{X}}) + h'(\underline{\underline{X}})\, g(\underline{\underline{X}}))(p(\underline{\underline{X}}) + h(\underline{\underline{X}})\, f(\underline{\underline{X}})) \in \mathfrak{P} .$$

However, since $\mathfrak{C}$ is closed under multiplication, $c'(\underline{\underline{X}})\, c(\underline{\underline{X}}) \in \mathfrak{C}$, contradicting the hypothesis that $\mathfrak{P} \cap \mathfrak{C} = \emptyset$.

Proof of Theorem 1D: Suppose $f \in \sqrt{\mathfrak{A}}$, so that there exists a positive integer m with $f^{m} \in \mathfrak{A}$. Thus for every $\underline{\underline{x}} \in A(\mathfrak{A})$, $f^{m}(\underline{\underline{x}}) = 0$. Hence $f(\underline{\underline{x}}) = 0$ for every $\underline{\underline{x}} \in A(\mathfrak{A})$. Therefore $f(\underline{\underline{X}}) \in \mathfrak{J}(A(\mathfrak{A}))$, and $\sqrt{\mathfrak{A}} \subseteq \mathfrak{J}(A(\mathfrak{A}))$.

Suppose $f \notin \sqrt{\mathfrak{A}}$. If $\mathfrak{C}$ is the set of all positive integer powers of f , then $\mathfrak{C} \cap \mathfrak{A} = \emptyset$; also $\mathfrak{C}$ does not contain zero. Let $\mathfrak{P}$ be an ideal containing $\mathfrak{A}$ which is maximal† with respect to the property that $\mathfrak{C} \cap \mathfrak{P} = \emptyset$. By Lemma 1F, $\mathfrak{P}$ is a prime ideal. By Lemma 1E, there exists a point $\underline{\underline{x}} \in \Omega^{n}$ such that $\mathfrak{P} = \mathfrak{J}(\underline{\underline{x}})$. Since $f \notin \mathfrak{P}$, $f(\underline{\underline{x}}) \neq 0$. Also, $(\overline{\underline{\underline{x}}}) = A(\mathfrak{J}(\underline{\underline{x}})) = A(\mathfrak{P}) \subseteq A(\mathfrak{A})$, so that $\underline{\underline{x}} \in A(\mathfrak{A})$. It follows that $f \notin \mathfrak{J}(A(\mathfrak{A}))$. Thus $\mathfrak{J}(A(\mathfrak{A})) \subseteq \sqrt{\mathfrak{A}}$.

---

†) The existence of such an ideal is guaranteed by Theorem 1A.

* the ideal generated by $\mathfrak{P}$ and $f(\underline{\underline{X}})$.

Suppose S is an algebraic set. We call S reducible if $S = S_1 \cup S_2$, where S_1, S_2 are algebraic sets, and $S \neq S_1, S_2$. Otherwise, we call S irreducible.

Example: Let $k = \mathbb{Q}$, $K = \mathbb{C}$, $n = 2$, and let $\mathfrak{J}$ be the ideal generated in $k[X_1, X_2]$ by the polynomial $f(X_1, X_2) = X_1^2 - X_2^2$. Then $S = A(\mathfrak{J})$ is the set of all $\underline{\underline{x}} \in C^2$ such that $x_1^2 - x_2^2 = 0$. If S_1 is the set of all $\underline{\underline{x}} \in \mathbb{C}^2$ with $x_1 + x_2 = 0$, and S_2 is the set of all $\underline{\underline{x}} \in \mathbb{C}^2$ with $x_1 - x_2 = 0$, then $S = S_1 \cup S_2$, and $S_1 \neq S \neq S_2$. Hence S is reducible.

THEOREM 1G. Let S be a non-empty algebraic set. The following four conditions are equivalent:

(i) $S = \overline{(\underline{\underline{x}})}$, i.e. S is the closure of a single point $\underline{\underline{x}}$,

(ii) S is irreducible,

(iii) $\mathfrak{J}(S)$ is a prime ideal in $k[\underline{\underline{X}}]$,

(iv) $S = A(\mathfrak{P})$, where $\mathfrak{P}$ is a prime ideal in $k[\underline{\underline{X}}]$.

Proof: (i) $\Rightarrow$ (ii). Suppose $S = A \cup B$, where A and B are algebraic sets, and $A \neq S \neq B$. We have $\underline{\underline{x}} \in S = A \cup B$. We may suppose that, say, $\underline{\underline{x}} \in A$. Then $S = \overline{(\underline{\underline{x}})} \subseteq \overline{A} = A$, whence $S = A$, which is a contradiction.

(ii) $\Rightarrow$ (iii). Suppose that $\mathfrak{J}(S)$ is not prime. Then we would have $f(\underline{\underline{X}})\, g(\underline{\underline{X}}) \in \mathfrak{J}(S)$ with neither $f(\underline{\underline{X}})$ nor $g(\underline{\underline{X}})$ in $\mathfrak{J}(S)$. Let $\mathfrak{A} = (\mathfrak{J}(S), f(\underline{\underline{X}}))$ (i.e. the ideal generated by $\mathfrak{J}(S)$ and $f(\underline{\underline{X}})$). Let $\mathfrak{B} = (\mathfrak{J}(S), g(\underline{\underline{X}}))$. Let $A = A(\mathfrak{A})$, $B = A(\mathfrak{B})$. In view of $S = A(\mathfrak{J}(S))$ and $\mathfrak{A} \supseteq \mathfrak{J}(S)$, we have $A \subseteq S$. But $A \neq S$ since $f \in \mathfrak{J}(A)$ and

$f \notin \mathfrak{J}(S)$. Thus $A \subsetneq S$. Similarly, $B \subsetneq S$. But we claim that $S = A \cup B$. Clearly $A \cup B \subseteq S$. On the other hand, if $\underline{\underline{x}} \in S$, then $f(\underline{\underline{x}})\, g(\underline{\underline{x}}) = 0$. Without loss of generality, let us assume that $f(\underline{\underline{x}}) = 0$. Then $\underline{\underline{x}}$ is a zero of every polynomial of $\mathfrak{A}$, so that $\underline{\underline{x}} \in A$. Therefore $S \subseteq A \cup B$. Thus $S = A \cup B$, with $A \neq S \neq B$. This contradicts the irreducibility of S .

(iii) ⇒ (iv). Set $\mathfrak{P} = \mathfrak{J}(S)$. Then $S = A(\mathfrak{J}(S)) = A(\mathfrak{P})$.

(iv) ⇒ (i). Choose $\underline{\underline{x}}$ according to Lemma 1E with $\mathfrak{J}(\underline{\underline{x}}) = \mathfrak{P}$. Then $S = A(\mathfrak{P}) = A(\mathfrak{J}(\underline{\underline{x}})) = (\overline{\underline{\underline{x}}})$. The proof of Theorem 1G is complete.

A set S satisfying any one of the four equivalent properties of Theorem 1G is called a <u>variety</u>. (More precisely, it is a k-<u>variety</u>.) If V is a variety, $\underline{\underline{x}} \in V$ is called a <u>generic point</u> of V if $V = (\overline{\underline{\underline{x}}})$.

<u>COROLLARY 1H</u>. <u>There is a one to one correspondence between the collection of all k - varieties V in Ω^n and the collection of all prime ideals $\mathfrak{P} \neq k[\underline{\underline{X}}]$ in $k[\underline{\underline{X}}]$, given by</u>

$$V \overset{\alpha}{\to} \mathfrak{P} = \mathfrak{J}(V) \quad \underline{\text{and}} \quad \mathfrak{P} \overset{\beta}{\to} V = A(\mathfrak{P}) .$$

<u>Proof</u>: Let V be a variety in Ω^n ; then $V \overset{\alpha}{\to} \mathfrak{J}(V) \overset{\beta}{\to} A(\mathfrak{J}(V)) = V$. Also, if $\mathfrak{P}$ is a prime ideal in $k[\underline{\underline{X}}]$, then $\mathfrak{P} \overset{\beta}{\to} A(\mathfrak{P}) \overset{\alpha}{\to} \mathfrak{J}(A(\mathfrak{P})) = \sqrt{\mathfrak{P}} = \mathfrak{P}$

<u>Examples</u>: (1) Let $S = \Omega^n$. Now $\mathfrak{J}(\Omega^n) = (0)$, a prime ideal. Suppose $\underline{\underline{x}} = (x_1,\ldots,x_n)$ is of transcendence degree n , i.e. the n

coordinates are algebraically independent over k. Then $\mathfrak{J}(\underline{x}) = (0)$, so $\overline{(\underline{x})} = A(\mathfrak{J}(\underline{x})) = A((0)) = \Omega^n$. So any point of Ω^n of transcendence degree n over k is a generic point of Ω^n.

(2) Let $k = \mathbb{Q}$, $\Omega = \mathbb{C}$, $n = 2$. Let $\mathfrak{P}$ be the principal ideal generated by $f(X_1,X_2) = X_1^2 + X_2^2 - 1$. $\mathfrak{P}$ is a prime ideal since f is irreducible. Thus $A(\mathfrak{P})$, i.e. the unit circle, is a variety. Choose $x_1 \in \Omega$ and transcendental over $\mathbb{Q}$. Pick $x_2 \in \Omega$ with $x_2^2 = 1 - x_1^2$. Then the point $\underline{x} = (x_1,x_2)$ belongs to $A(\mathfrak{P})$. In fact, $\underline{x}$ is a generic point of $A(\mathfrak{P})$:

To see this, it will suffice to show that $\mathfrak{J}(\underline{x}) = (X_1^2 + X_2^2 - 1)$, i.e. the principal ideal generated by $X_1^2 + X_2^2 - 1$. If $g(X_1,X_2) \in \mathfrak{J}(\underline{x})$, that is, if $g(x_1,x_2) = 0$, then $g(x_1,X_2)$ is a multiple of $X_2^2 - 1 + x_1^2$, since x_2 is a root of $X_2^2 - 1 + x_1^2$, which is irreducible over $\mathbb{Q}(x_1)$. More precisely,

$$g(x_1,X_2) = (X_2^2 - 1 + x_1^2)\, h(x_1,X_2) ,$$

where $h(X_1,X_2)$ is a polynomial in X_2 and is rational in X_1. Since x_1 was transcendental, we get

$$g(X_1,X_2) = (X_1^2 + X_2^2 - 1)\, h(X_1,X_2) .$$

In view of the unique factorization in $\mathbb{Q}[X_1]$, it follows that $h(X_1,X_2)$ is in fact a polynomial in X_1,X_2. Thus $\mathfrak{J}(\underline{x}) = (X_1^2 + X_2^2 - 1)$.

(3) Let $k = \mathbb{Q}$, $\Omega = \mathbb{C}$, $n = 2$. Let $\mathfrak{P}$ be the principal ideal generated by $f(X_1,X_2) = X_1^2 - X_2$. Then $A(\mathfrak{P})$ is irreducible and is a parabola. Choose $x_1 \in \Omega$ and transcendental over Q, and put $x_2 = x_1^2$. Then $\underline{x} = (x_1,x_2)$ lies in $A(\mathfrak{P})$. An argument similar to

the one given in (2) shows that $\underline{\underline{x}}$ is a generic point of $A(\mathfrak{P})$. For example, Lindemann's Theorem says that e is transcental over $\mathbb{Q}$, and therefore (e,e^2) is a generic point of $A(\mathfrak{P})$.

(4) Let $k = \mathbb{Q}$, $\Omega = \mathbb{C}$. Let $\mathfrak{U}$ be the principal ideal $\mathfrak{U} = (X_1^2 - X_2^2)$. Then as we have seen above, $A(\mathfrak{U})$ is reducible and is therefore not a variety.

(5) Consider a linear manifold M^d given by a parameter representation

$$x_i = b_i + a_{i1} t_1 + \dots + a_{id} t_d \qquad (1 \leq i \leq n) .$$

Here the b_i and the a_{ij} as given elements of k, with the $(d \times n)$ - matrix (a_{ij}) of rank d . As $t_1,\dots,t_d$ run through Ω , $\underline{\underline{x}} = (x_1,\dots,x_n)$ runs through M^d . It follows from linear algebra that M^d is an algebraic set. (It is a "d-dimensional linear manifold". See also §2 about the notion of dimension). In fact M^d is a variety:

Choose $\eta_1,\dots,\eta_d$ algebraically independent over k . Put

$$\xi_i = b_i + a_{i1} \eta_1 + \dots + a_{id} \eta_d \quad (1 \leq i \leq n)$$

and $\underline{\underline{\xi}} = (\xi_1,\xi_2,\dots,\xi_n) \in \Omega^n$. Now $\underline{\underline{\xi}} \in M^d$, so $(\underline{\underline{\xi}}) \subseteq M^d$. Conversely, if $f(\underline{\underline{\xi}}) = 0$, then

$$f(b_1 + a_{11}T_1 + \dots + a_{1d}T_d ,$$
$$b_2 + a_{21}T_1 + \dots + a_{2d}T_d,\dots,b_n + a_{n1}T_1 + \dots + a_{nd}T_d) = 0 ,$$

where $T_1,\dots,T_d$ are variables. Thus if $\underline{\underline{x}} \in M^d$, then $f(\underline{\underline{x}}) = 0$. So every $\underline{\underline{x}} \in M^d$ lies in $A(\mathfrak{J}(\underline{\underline{\xi}})) = (\underline{\underline{\xi}})$. Therefore we have shown that $M^d = (\underline{\underline{\xi}})$, or that M^d is a variety.

(6) Take $k = \mathbb{Q}$, $\Omega = \mathbb{C}$, $n = 2$, and $\mathfrak{A}$ the principal ideal generated by $f(X_1,X_2) = X_1^2 - 2X_2^2$. Over $k = \mathbb{Q}$, this polynomial is irreducible. Thus $\mathfrak{A}$ is a prime ideal, and $A(\mathfrak{A})$ is a variety. However, if we take $k' = \mathbb{Q}(\sqrt{2})$, then $f(X_1,X_2)$ is no longer irreducible over k', so that $\mathfrak{A}$ is no longer a prime ideal in $k'[X_1,X_2]$, and $A(\mathfrak{A})$ is no longer a variety.

This prompts the definition: A variety is called an <u>absolute variety</u> if it remains a variety over every algebraic extension of k.

<u>THEOREM 1I. Every non-empty algebraic set is a finite union of varieties.</u>

<u>Proof</u>: We first show that every non-empty collection $\mathfrak{S}$ of algebraic sets has a minimal element. For if we form all ideals $\mathfrak{J}(S)$, where $S \in \mathfrak{S}$, there is by Theorem 1A a maximal element of this non-empty collection of ideals. Say $\mathfrak{J}(S_0)$ is maximal. We claim that $S_0 \in \mathfrak{S}$ is minimal. For if $S_1 \subseteq S_0$ where $S_1 \in \mathfrak{S}$, then $\mathfrak{J}(S_1) \supseteq \mathfrak{J}(S_0)$; but since $\mathfrak{J}(S_0)$ is maximal, $\mathfrak{J}(S_1) = \mathfrak{J}(S_0)$. Thus $S_1 = A(\mathfrak{J}(S_1))$ $= A(\mathfrak{J}(S_0)) = S_0$.

Suppose that Theorem 1I is false. Let $\mathfrak{S}$ be the collection of algebraic sets for which Theorem 1I is false. There is a minimal element S_0 of $\mathfrak{S}$. If S_0 were a variety, then the theorem would be true for S_0. Hence S_0 is reducible. Let $S_0 = A \cup B$, where A,B are algebraic sets, with $A \neq S_0 \neq B$. Since S_0 is minimal and $A \subsetneqq S_0$, $B \subsetneqq S_0$, the theorem is true for A,B. Hence, we can write $A = V_1 \cup \ldots \cup V_m$, and $B = W_1 \cup \ldots \cup W_\ell$, where $V_i\,(1 \le i \le m)$ and $W_j\ (1 \le j \le \ell)$ are varieties. Thus

$$S_0 = A \cup B = V_1 \cup \ldots \cup V_m \cup W_1 \cup \ldots \cup W_\ell ,$$

contradicting our hypothesis that $S_0 \in \mathfrak{C}$.

It is clear that there exists a representation of S as $S = V_1 \cup \ldots \cup V_t$ where $V_i \not\subseteq V_j$ if $i \neq j$.

THEOREM 1J. *Let S be a non-empty algebraic set. The representation of S as*

$$S = V_1 \cup \ldots \cup V_t ,$$

where $V_1, \ldots, V_t$ are varieties with $V_i \not\subseteq V_j$ if $i \neq j$, is unique.

Proof: Exercise.

The V_i in the unique representation of S given in Theorem 1J are called the *components* of S .

Example: Let $k = \mathbb{Q}$, $\Omega = \mathbb{C}$, $n = 2$, and $S = A((X_1^2 - X_2^2))$. Let $V_1 = A((X_1 - X_2))$ and $V_2 = A((X_1 + X_2))$; then $S = V_1 \cup V_2$. Here V_1, V_2 are two intersecting lines.

Finally we introduce the following terminology and notation. We say $\underline{\underline{y}}$ is a *specialization* of $\underline{\underline{x}}$ and write

$$\underline{\underline{x}} \to \underline{\underline{y}} ,$$

if $\underline{\underline{y}} \in (\overline{\underline{\underline{x}}})$. This holds precisely if $f(\underline{\underline{y}}) = 0$ for every $f(\underline{\underline{X}}) \in k[\underline{\underline{X}}]$ with $f(\underline{\underline{x}}) = 0$. It is immediately seen that $\to$ is transitive, i.e. that

$$\underline{\underline{x}} \to \underline{\underline{y}} \text{ and } \underline{\underline{y}} \to \underline{\underline{z}} \text{ implies that } \underline{\underline{x}} \to \underline{\underline{z}} .$$

If both $\underline{\underline{x}} \to \underline{\underline{y}}$ and $\underline{\underline{y}} \to \underline{\underline{x}}$, then we write $\underline{\underline{x}} \leftrightarrow \underline{\underline{y}}$. This is equivalent with the equation $(\bar{\underline{x}}) = (\bar{\underline{y}})$.

Example: Let $\underline{x} = (e,e^2)$ and $\underline{y} = (1,1)$. Then $\underline{x} \to \underline{y}$. For as we saw in example (3) below Theorem 1G , the point $\underline{x}$ is a generic point of the parabola $x_2 - x_1^2 = 0$, and $\underline{y}$ lies on this parabola.

§2. Dimension.

Let $\underline{x} \in \Omega^n$. The transcendence degree of $\underline{x}$ over k is the maximum number of algebraically independent components of $\underline{x}$ over k . This clearly is equal to the transcendence degree of $k(\underline{x})$ over k . We have

$$0 \le \text{tr. deg. } \underline{x} \le n .$$

THEOREM 2A. Suppose $\underline{x} \to \underline{y}$. Then

(i) tr. deg. $\underline{y}$ $\le$ tr. deg. $\underline{x}$.

(ii) Equality hold in (i) if and only if $\underline{x} \leftrightarrow \underline{y}$.

Proof: (i) Induction on n . If $n = 1$, and if trans. deg. $\underline{x} = 1$, then tr. deg. $\underline{y} \le n = 1$ = trans. deg $\underline{x}$; if tr. deg. $\underline{x} = 0$, then $\underline{x}$ is algebraic over k . In this case, since $\underline{x} \to \underline{y}$, the components of $\underline{y}$ satisfy the algebraic equations satisfied by the components of $\underline{x}$, and tr. deg. $\underline{y} = 0$.

To show the induction step, let d be the transcendence degree of $\underline{x}$. We may assume that $d < n$. We may also assume that tr. deg. $\underline{y} \ge d$. Without loss of generality, we assume that $y_1,\dots,y_d$ are algebraically independent over k . Since $\underline{x} = (x_1,\dots,x_n) \to (y_1,\dots,y_n) = \underline{y}$, it follows that $(x_1,\dots,x_d) \to (y_1,\dots,y_d)$. By induction, and since

$d < n$, the elements $x_1,\dots,x_d$ are also algebraically independent over k . Let $d < i \le n$. Then x_i is algebraically dependent on $x_1,\dots,x_d$. So x_i satisfies some non-trivial equation

$$x_i^a\, g_a(x_1,\dots,x_d) + x_i^{a-1}\, g_{a-1}(x_1,\dots,x_d) + \dots + g_0(x_1,\dots,x_d) = 0$$

Since $\underline{\underline{x}} \to \underline{\underline{y}}$, it follows that

$$y_i^a\, g_a(y_1,\dots,y_d) + y_i^{a-1}\, g_a(y_1,\dots,y_d) + \dots + g_0(y_1,\dots,y_d) = 0 .$$

Thus y_i is algebraically dependent on $y_1,\dots,y_d$. This is true for any i in $d < i \le n$. So tr. deg. $\underline{\underline{y}} \le d$.

(ii) If $\underline{\underline{x}} \leftrightarrow \underline{\underline{y}}$, then it follows from part (i) that tr. deg. $\underline{\underline{x}}$ = tr. deg. $\underline{\underline{y}}$.

Suppose $\underline{\underline{x}} \to \underline{\underline{y}}$ and tr. deg. $\underline{\underline{x}}$ = tr. deg. $\underline{\underline{y}}$. Let the common transcendence degree be d . We may assume without loss of generality that the first d coordinates $y_1,\dots,y_d$ are algebraically independent over k . Then by part (i) and by $(x_1,\dots,x_d) \to (y_1,\dots,y_d)$, also $x_1,\dots,x_d$ are algebraically independent over k . We have to show that $\underline{\underline{y}} \to \underline{\underline{x}}$, i.e. that if $f(\underline{\underline{y}}) = 0$ for $f \in k[\underline{\underline{X}}]$, then $f(\underline{\underline{x}}) = 0$. Put differently, we have to show that if $f(\underline{\underline{x}}) \ne 0$, then $f(\underline{\underline{y}}) \ne 0$. So let $f(\underline{\underline{x}}) \ne 0$. Then $f(\underline{\underline{x}})$ is a non-zero element of $k(\underline{\underline{x}})$ and $1/f(\underline{\underline{x}}) \in k(\underline{\underline{x}})$. Now since $x_{d+1},\dots,x_n$ are algebraic over $k(x_1,\dots,x_d)$, it is well known that

$$k(\underline{\underline{x}}) = k(x_1,\dots,x_d)[x_{d+1},\dots,x_n] ,$$

i.e. $k(\underline{\underline{x}})$ is obtained from $k(x_1,\dots,x_d)$ by forming the polynomial ring in $x_{d+1},\dots,x_n$.

Thus

$$1/f(\underline{\underline{x}}) = v(x_1,\ldots,x_n)/u(x_1,\ldots,x_d) ,$$

where $v(X_1,\ldots,X_n)$ and $u(X_1,\ldots,X_d)$ are polynomials. We have

$$u(x_1,\ldots,x_d) = f(\underline{\underline{x}})\ v(\underline{\underline{x}}) ,$$

which implies that

$$u(y_1,\ldots,y_d) = f(\underline{\underline{y}})\ v(\underline{\underline{y}}) ,$$

in view of $\underline{\underline{x}} \to \underline{\underline{y}}$. Now $y_1,\ldots,y_d$ are independent over k , whence $u(y_1,\ldots,y_d) \neq 0$, whence $f(\underline{\underline{y}}) \neq 0$. Our proof is complete.

The _dimension_ of a variety V is defined as the transcendence degree of any of its generic points. In view of Theorem 2A , there is no ambiguity. A variety of dimension 1 is called a _curve_, one of dimension $n - 1$ is called a _hypersurface_.

Example: Let us consider again the example of the linear manifold M^d . We constructed a generic point $(\xi_1,\ldots,\xi_n)$ with $k(\eta_1,\ldots,\eta_d) = k(\xi_1,\ldots,\xi_n)$, where $\eta_1,\ldots,\eta_d$ were algebraically independent. Thus tr. deg. $k(\xi_1,\ldots,\xi_n) = d$. Hence in the sense of our definition, M^d has dimension d . This agrees with the dimension d assigned to M^d in linear algebra.

THEOREM 2B. (i) _Let_ V _be a variety and let_ $\underline{\underline{x}} \in V$ _with tr. deg._ $\underline{\underline{x}} = \dim V$. _Then_ $\underline{\underline{x}}$ _is a generic point of_ V .

(ii) _If_ $W \subseteq V$ _are two varieties, and if_ $\dim W = \dim V$, _then_ $W = V$.

Proof: (i) Let $\mathbf{y}$ be a generic point of V. Then $\mathbf{y} \to \mathbf{x}$ and tr. deg. $\mathbf{x}$ = tr. deg. $\mathbf{y}$. By Theorem 2A, $\mathbf{x} \leftrightarrow \mathbf{y}$, so that $\overline{(\mathbf{x})} = \overline{(\mathbf{y})} = V$.

(ii) Let $\mathbf{x}$ be a generic point of W. Now $\mathbf{x} \in V$, and tr. deg. $\mathbf{x} = \dim V$, so that by part (i), $\mathbf{x}$ is a generic point of V. Thus $\overline{(\mathbf{x})} = W = V$.

THEOREM 2C. (i) *If $f(\mathbf{X}) \in k[\mathbf{X}]$ is a non-constant irreducible polynomial, then the set of zeros of $f(\mathbf{X})$ is a hypersurface; that is, a variety of dimension $n-1$.*

(ii) *If S is a hypersurface, then $\mathfrak{J}(S)$ is a principal ideal (f), generated by some non-constant irreducible polynomial $f(\mathbf{X}) \in k[\mathbf{X}]$.*

Proof: (i) The principal ideal (f) is a prime ideal in $k[\mathbf{X}]$, so $A((f))$ is a variety. Without loss of generality, suppose X_n occurs in $f(\mathbf{X})$, say $f(\mathbf{X}) = X_n^a g_a(X_1,\ldots,X_{n-1}) + \ldots + g_0(X_1,\ldots,X_{n-1})$. Choose $x_1,\ldots,x_{n-1} \in \Omega$ algebraically independent over k. Choose $x_n \in \Omega$ with $f(x_1,\ldots,x_n) = 0$. Then $\mathbf{x} = (x_1,\ldots,x_n) \in A((f))$. Also, tr. deg. $\mathbf{x} = n-1$. Thus $\dim A((f)) \geq n-1$. On the other hand, $\dim A((f)) \neq n$, by Theorem 2B and since $A((f)) \neq \Omega^n$. Hence $\dim A((f)) = n-1$. In other words, $A((f))$ is a hypersurface.

(ii) If S is a hypersurface, then $\mathfrak{J}(S)$ is a prime ideal. Let $g(\mathbf{X}) \in \mathfrak{J}(S)$, $g \neq 0$. Since $\mathfrak{J}(S)$ is prime, there exists some irreducible factor f of g such that $f(\mathbf{X}) \in \mathfrak{J}(S)$. So $(f) \subseteq \mathfrak{J}(S)$, whence $A((f)) \supseteq A(\mathfrak{J}(S)) = S$. But $\dim A((f)) = n-1$ by part (i), and $\dim S = n-1$. Therefore by Theorem 2B, $A(f) = S$. Hence

$$\mathfrak{J}(S) = \mathfrak{J}(A(f)) = \sqrt{(f)} = (f),$$

since (f) is prime.

Examples: (1) Let $k = \mathbb{Q}$, $\Omega = \mathbb{C}$, $n = 2$ and $f(X,Y) = Y - X^2$. Now f is irreducible. So by Theorem 2C , the set of zeros of f is a hypersurface of dimension 1 . Since $n - 1 = 1$, it is also a curve. The point (e, e^2) has transcendence degree 1 and lies on our curve. Hence we see again that it is a generic point of our curve.

(2) Same as above, but with $f(X,Y) = X^2 + Y^2 - 1$. Again the set of zeros of f (namely the unit circle) is a hypersurface and also a curve.

Let t be transcendental and consider the point

$$\underline{\underline{x}} = (x_1, x_2) = \left(\frac{2t}{t^2+1} , \frac{t^2-1}{t^2+1} \right) .$$

Here $t = \frac{x_1}{1-x_2}$, whence $k(\underline{\underline{x}}) = k(t)$, so that $\underline{\underline{x}}$ has transcendence degree 1 . Since $\underline{\underline{x}}$ lies on our curve, it follows that $\underline{\underline{x}}$ is a generic point of the unit circle. In particular,

$$\left(\frac{2e}{e^2+1} , \frac{e^2-1}{e^2+1} \right)$$

is a generic point of the unit circle.

THEOREM 2D. Let $n = 1 + t$, let $f_1(X, Y_1)$, $f_2(X, Y_1, Y_2), \ldots, f_t(X, Y_1, Y_2, \ldots, Y_t)$ be polynomials of the type

$$f_i(X, Y_1, \ldots, Y_i) = Y_i^{d_i} - g_i(X, Y_1, \ldots, Y_i) ,$$

where $d_i > 0$ and g_i is of degree $< d_i$ in Y_i . Let $\mathfrak{Y}_1, \ldots, \mathfrak{Y}_t$ be algebraic functions with $f_1(X, \mathfrak{Y}_1) = \ldots = f_t(X, \mathfrak{Y}_1, \ldots, \mathfrak{Y}_t) = 0$, and suppose that

$$[k(X,\mathfrak{Y}_1,\ldots,\mathfrak{Y}_t) : k(X)] = d_1 d_2 \ldots d_t .$$

<u>Then the equations</u>

$$f_1 = f_2 = \ldots = f_t = 0$$

<u>define a curve; that is, a variety of dimension</u> 1 .

<u>Examples</u>: (1) Let k be a field whose characteristic does not equal 2 or 3 . Take $t = 2$, so that $n = 3$. Consider $f_1(X,Y_1) = Y_1^2 + X^2 - 1$, $f_2(X,Y_1Y_2) = Y_2^2 + X^2 - 4$. Then $\mathfrak{Y}_1^2 = 1 - X^2$, and $\mathfrak{Y}_2^2 = 4 - X^2$, or $\mathfrak{Y}_1 = \sqrt{1 - X^2}$ and $\mathfrak{Y}_2 = \sqrt{4 - X^2}$. Also,

(2.1) $$[k(X,\sqrt{1 - X^2}, \sqrt{4 - X^2}) : k(X)] = 4 .^{\dagger)}$$

By Theorem 2D , the equations $f_1 = f_2 = 0$ define a curve. This curve is the intersection of two circular cylinders with radii 1,2 , whose axes intersect at right angles.

(2) Same as above, but with $f_2(X,Y_1,Y_2) = Y_2^2 + X^2 - 1$. In this case $[k(X,\mathfrak{Y}_1,\mathfrak{Y}_2) : k(X)] = 2$. So Theorem 2D does not apply. In fact,

---

†) The proof of (2.1) is as follows. Since the characteristic is not 2 or 3 , the four polynomials $1 - X$, $1 + X$, $2 - X$, $2 + X$ are distinct and are irreducible. Hence none of $1 - X^2$, $4 - X^2$ and $(1 - X^2)/(4 - X^2)$ is a square in $k(X)$, and each of $\sqrt{1 - X^2}$, $\sqrt{4 - X^2}$, $\sqrt{(1 - X^2)/(4 - X^2)}$ is of degree 2 over $k(X)$. It will suffice to show that $\sqrt{4 - X^2} \notin k(X,\sqrt{1 - X^2})$. Suppose to the contrary that

$$\sqrt{4 - X^2} = r(X) + s(X)\sqrt{1 - X^2}$$

with rational functions $r(X)$, $s(X)$. We now square and observe that the factor in front of $\sqrt{1 - X^2}$ must be zero. Thus $2r(X)\,s(X) = 0$. If $r(X) = 0$, then $(1 - X^2)/(1 - X^4)$ would be a square in $k(X)$, which was ruled out. If $s(X) = 0$, then $4 - X^2$ would be a square, which was also ruled out.

The situation is similar to the one in Corollary 5B of Chapter II, §5, and the exercise below it.

$$A((f_1, f_2)) = V_1 \cup V_2 ,$$

where $V_1 = A((f_1, Y_1 - Y_2))$, $V_2 = A(f_1, Y_1 + Y_2))$,
Thus we do not obtain a variety. This algebraic set is the intersection of two circular cylinders of radius 1 whose axes intersect at right angles. Both V_1 and V_2 are the intersection of a plane with a circular cylinder; they are ellipses.

(3) Let $k = F_q$, the finite field of q elements. Take $t = 2$, $n = 3$ and $f_1(X, Y_1) = Y_1^d - f(X)$ where $d|(q-1)$, and $f_2(X, Y_2) = Y_2^q - Y_2 - g(X)$. Suppose f_1, f_2 to be irreducible. Then $\mathfrak{Y}_1, \mathfrak{Y}_2$ with $\mathfrak{Y}_1^d = f(X)$, $\mathfrak{Y}_2^q - \mathfrak{Y}_2 = g(X)$ have

$$[k(X, \mathfrak{Y}_1) : k(X)] = d , \quad [k(X, \mathfrak{Y}_2) : k(X)] = q .$$

Since $(d,q) = 1$, we have $[k(X, \mathfrak{Y}_1, \mathfrak{Y}_2) : k(X)] = dq$. Thus $f_1 = f_2 = 0$ defines a curve. In the same way one sees that if f_1, f_2 both are absolutely irreducible, then $f_1 = f_2 = 0$ is an absolute curve, i.e., a curve which is an <u>absolute</u> variety.

<u>Proof of Theorem 2D:</u> Pick $\underline{\underline{x}} = (x, y_1, \ldots, y_t) \in \Omega^n$, such that the mapping $X \to x$, $\mathfrak{Y}_i \to y_i$ $(1 \le i \le t)$ yields an isomorphism of $k(X, \mathfrak{Y}_1, \ldots, \mathfrak{Y}_t)$ to $k(x, y_1, \ldots, y_t)$. We claim that the set of zeros of $f_1 = f_1 = \ldots = f_t = 0$ is the variety $(\overline{\underline{\underline{x}}})$. It suffices to show that $\mathfrak{J}(\underline{\underline{x}}) = (f_1, \ldots, f_t)$; for then $(\overline{\underline{\underline{x}}}) = A(\mathfrak{J}(\underline{\underline{x}})) = A((f_1, \ldots, f_t))$. Clearly, every $f \in (f_1, \ldots, f_t)$ vanishes on $\underline{\underline{x}}$; so $(f_1, \ldots, f_t) \subseteq \mathfrak{J}(\underline{\underline{x}})$. Conversely, we are going to show that

(2.2) <u>if</u> $f(\underline{\underline{x}}) = 0$, <u>then</u> $f \in (f_1, \ldots, f_t)$.

We'll show (2.2) by induction on s, for functions $f = f(X, Y_1, \ldots, Y_s)$ where $0 \leqq s \leqq t$. If $s = 0$, then $f(x) = 0$; but x is transcendental over k, so $f(X) = 0$, whence $f \in (f_1, \ldots, f_t)$. Next, we show that if (2.2) is true for $s-1$, it is true for s. In $f(X, Y_1, \ldots, Y_s)$, if $Y_s^{d_s}$ occurs, replace it by $g_s(X, Y_1, \ldots, Y_s)$. Do this repeatedly, until you get a polynomial $\hat{f}(X, Y_1, \ldots, Y_s)$ of degree $< d_s$ in Y_s. We observe that $f - \hat{f} \in (f_s)$, and that $\hat{f}(\underline{\underline{x}}) = 0$. Suppose

$$\hat{f} = Y_s^{d_s - 1} h_{d_s - 1}(X, Y_1, \ldots, Y_{s-1}) + \ldots + h_0(X, Y_1, \ldots, Y_{s-1}). \tag{2.3}$$

Our hypothesis implies that $[k(x, y_1, \ldots, y_t) : k(x)] = d_1 d_2 \ldots d_t$, and we have

$$k(x) \subseteq k(x, y_1) \subseteq k(x, y_1, y_2) \subseteq \ldots \subseteq k(x, y_1, \ldots, y_t),$$

where for each i in $1 \leq i \leq t$, the field $k(x, y_1, \ldots, y_i)$ is an extension of degree $\leq d_i$ over $k(x, y_1, \ldots, y_{i-1})$. Hence it is actually an extension of degree d_i. In particular, $[k(x, y_1, \ldots, y_s) : k(x, y_1, \ldots, y_{s-1})] = d_s$. Since $\hat{f}(\underline{\underline{x}}) = 0$, we see from (2.3) that each $h_j(\underline{\underline{x}}) = 0$. So by induction, each $h_j \in (f_1, \ldots, f_t)$, hence also $\hat{f} \in (f_1, \ldots, f_t)$, and $f \in (f_1, \ldots, f_t)$. The proof of (2.2) and therefore the proof of the theorem is complete.

§3. Rational Maps.

A rational function φ on Ω^n is an element of $k(X_1, \ldots, X_n)$, i.e. of the form $\varphi = a(X_1, \ldots, X_n)/b(X_1, \ldots, X_n)$, where $a(X_1, \ldots, X_n)$, $b(X_1, \ldots, X_n)$ are polynomials over k. We may assume that a, b have no common factor.

We say a rational function φ is <u>defined</u> (or <u>regular</u>) at a point $\underline{\underline{x}} \in \Omega^n$ if $b(\underline{\underline{x}}) \neq 0$. If φ is defined at $\underline{\underline{x}}$, put $\varphi(\underline{\underline{x}}) = a(\underline{\underline{x}})/b(\underline{\underline{x}})$.

The rational functions φ which are defined at $\underline{\underline{x}} \in \Omega^n$ form a ring consisting of all $a(\underline{\underline{X}})/b(\underline{\underline{X}})$ with $b(\underline{\underline{x}}) \neq 0$. This ring is denoted as $\mathfrak{O}_{\underline{\underline{x}}}$ and is called the <u>local ring</u> of $\underline{\underline{x}}$. Let $\mathfrak{I}_{\underline{\underline{x}}}$ consist of all $\varphi \in \mathfrak{O}_{\underline{\underline{x}}}$ with $\varphi(\underline{\underline{x}}) = 0$. (Thus $\mathfrak{I}_{\underline{\underline{x}}}$ consists of all $a(\underline{\underline{X}})/b(\underline{\underline{X}})$ with $b(\underline{\underline{x}}) \neq 0$, $a(\underline{\underline{x}}) = 0$.) Then $\mathfrak{I}_{\underline{\underline{x}}}$ is an ideal in $\mathfrak{O}_{\underline{\underline{x}}}$.

<u>LEMMA 3A</u>. (i) <u>If</u> $\underline{\underline{x}} \to \underline{\underline{y}}$, <u>then</u> $\mathfrak{O}_{\underline{\underline{y}}} \subseteq \mathfrak{O}_{\underline{\underline{x}}}$.

(ii) <u>If</u> $\underline{\underline{x}} \leftrightarrow \underline{\underline{y}}$, <u>then</u> $\mathfrak{O}_{\underline{\underline{x}}} = \mathfrak{O}_{\underline{\underline{y}}}$ <u>and</u> $\mathfrak{I}_{\underline{\underline{x}}} = \mathfrak{I}_{\underline{\underline{y}}}$.

<u>Proof</u>: Obvious.

<u>THEOREM 3B</u>. (i) $\mathfrak{I}_{\underline{\underline{x}}}$ <u>is a maximal ideal in</u> $\mathfrak{O}_{\underline{\underline{x}}}$, <u>hence</u> $\mathfrak{O}_{\underline{\underline{x}}}/\mathfrak{I}_{\underline{\underline{x}}}$ <u>is a field (called the function field of</u> $\underline{\underline{x}}$).

(ii) $\mathfrak{O}_{\underline{\underline{x}}}/\mathfrak{I}_{\underline{\underline{x}}}$ <u>is</u> k-<u>isomorphic to</u> $k(\underline{\underline{x}})$.

<u>Proof</u>: (i) Let $\varphi \in \mathfrak{O}_{\underline{\underline{x}}}$, $\varphi \notin \mathfrak{I}_{\underline{\underline{x}}}$. Then $\varphi = a(\underline{\underline{X}})/b(\underline{\underline{X}})$, where $b(\underline{\underline{x}}) \neq 0$ and $a(\underline{\underline{x}}) \neq 0$, and therefore $\frac{1}{\varphi} = b(\underline{\underline{X}})/a(\underline{\underline{X}})$ lies in $\mathfrak{O}_{\underline{\underline{x}}}$. Thus every $\varphi \in \mathfrak{O}_{\underline{\underline{x}}}$ which does not lie in $\mathfrak{I}_{\underline{\underline{x}}}$ is a unit. It follows that $\mathfrak{I}_{\underline{\underline{x}}}$ is a maximal ideal.

(ii) The map $\omega\colon \mathfrak{O}_{\underline{\underline{x}}} \to k(\underline{\underline{x}})$ given by

$$\omega(a(\underline{\underline{X}})/b(\underline{\underline{X}})) = a(\underline{\underline{x}})/b(\underline{\underline{x}})$$

has image $k(\underline{\underline{x}})$ and kernel $\mathfrak{I}_{\underline{\underline{x}}}$. Therefore $k(\underline{\underline{x}}) \cong \mathfrak{O}_{\underline{\underline{x}}}/\mathfrak{I}_{\underline{\underline{x}}}$.

We now come to the definition of a rational function defined on a variety V. The simplest definition to try would be that a rational function on V is the restriction to V of a rational function $\varphi(\underline{\underline{X}})$ on Ω^n. However, we want this rational function to be defined for at least some point of V. Hence by Lemma 3A it must be defined for every generic point $\underline{\underline{x}}$ of V, i.e. it must lie in $\mathfrak{O}_{\underline{\underline{x}}}$. Moreover, given two functions $a(\underline{\underline{X}})/b(\underline{\underline{X}})$ and $c(\underline{\underline{X}})/d(\underline{\underline{X}})$ in $\mathfrak{O}_{\underline{\underline{x}}}$, we should regard them as equal functions on V if their restrictions to V are equal. Clearly this is true precisely if their difference lies in $\mathfrak{J}_{\underline{\underline{x}}}$.

Thus we come to define a <u>rational function on</u> V as an element of $\mathfrak{O}_{\underline{\underline{x}}}/\mathfrak{J}_{\underline{\underline{x}}}$, where $\underline{\underline{x}}$ is a generic point. Clearly this is independent of the choice of the generic point. $\mathfrak{O}_{\underline{\underline{x}}} = \mathfrak{O}_V$ (say) consists of $a(\underline{\underline{X}})/b(\underline{\underline{X}})$ with $b(\underline{\underline{X}}) \notin \mathfrak{J}(V) = \mathfrak{J}(\underline{\underline{x}})$, and $\mathfrak{J}_{\underline{\underline{x}}} = \mathfrak{J}_V$ (say) consists of $a(\underline{\underline{X}})/b(\underline{\underline{X}})$ with $a(\underline{\underline{X}}) \in \mathfrak{J}(V)$, $b(\underline{\underline{X}}) \notin \mathfrak{J}(V)$. We say a function $r(\underline{\underline{X}}) \in k(\underline{\underline{X}})$ <u>represents</u> a rational function φ of V if $r(\underline{\underline{X}}) \in \mathfrak{O}_V$ and if $r(\underline{\underline{X}})$ lies in the class φ of $\mathfrak{O}_V/\mathfrak{J}_V$.

<u>Example</u>: Let $n = 2$, $k = \mathbb{Q}$, $\Omega = \mathbb{C}$, and V the circle $x_1^2 + x_2^2 - 1 = 0$. Let φ be the rational function represented by X_1/X_2. Then φ is also represented by $(X_1 + X_1^2 + X_2^2 - 1)/X_2$ and by $X_1/(X_2 + X_1^2 + X_2^2 - 1)$, for example.

The rational functions defined on V form a field, called the <u>function field</u> of V. This field is denoted $k(V)$. In view of Theorem 3B, the function field is k-isomorphic to $k(\underline{\underline{x}})$ where $\underline{\underline{x}}$ is any generic point of V.

Let $\psi_1^V, \dots, \psi_n^V$ be the elements of $k(V)$ represented, respectively, by the polynomials $X_1, \dots, X_n$. Then it is clear that

$$k(V) = k(\psi_1^V, \dots, \psi_n^V) .$$

It is easily seen that a polynomial $f(X_1,\dots,X_n)$ has $f(\overset{V}{\psi}_1,\dots,\overset{V}{\psi}_n) = 0$ if and only if $f \in \mathfrak{J}(V)$. Hence if $\underline{\underline{x}} = (x_1,\dots,x_n)$ is a generic point, then there is a k-isomorphism $k(\underline{\underline{x}}) \to k(V)$ with $x_i \to \overset{V}{\psi}_i$ $(i = 1,\dots,n)$.

Example: Let $n = 2$, $k = \mathbb{Q}$, $\Omega = \mathbb{C}$, and V the circle $x_1^2 + x_2^2 - 1 = 0$. We have seen in previous examples that if η is trancendental over $\mathbb{Q}$, then the point $\left(2\eta/(\eta^2+1),\,(\eta^2-1)/(\eta^2+1)\right)$ is a generic point for V . Clearly $k(\underline{\underline{x}}) = k(\eta) \cong k(X)$. Thus the function field of the circle is isomorphic to $k(X)$.

A curve is called rational if its function field is $\cong k(X)$. Thus the circle is a rational curve. It can be shown that $x_1^n + x_2^n - 1 = 0$ is not a rational curve if $n > 2$ and is not divisible by the characteristic. See Shafarevich (1969), p. 8 .

Let φ be a rational function on a variety $V = \overline{(\underline{\underline{x}})}$ and let $\underline{\underline{y}}$ be a point of V . We say that φ is defined at $\underline{\underline{y}}$ if there exists a representative $r(\underline{\underline{X}}) = a(\underline{\underline{X}})/b(\underline{\underline{X}})$ with $b(\underline{\underline{y}}) \neq 0$. If this is the case, set

$$\varphi(\underline{\underline{y}}) = a(\underline{\underline{y}})/b(\underline{\underline{y}}) .$$

We have to show that this independent of the representative. Suppose that φ is represented by both $a(\underline{\underline{X}})/b(\underline{\underline{X}})$ and by $\hat{a}(\underline{\underline{X}})/\hat{b}(\underline{\underline{X}})$, and that $b(\underline{\underline{y}}) \neq 0$, $\hat{b}(\underline{\underline{y}}) \neq 0$. The difference $(a\hat{b} - \hat{a}b)/(b\hat{b})$ represents the zero rational function on V . Hence $a(\underline{\underline{x}})\hat{b}(\underline{\underline{x}}) - \hat{a}(\underline{\underline{x}})b(\underline{\underline{x}}) = 0$, and since $\underline{\underline{x}} \to \underline{\underline{y}}$, we have $a(\underline{\underline{y}})\hat{b}(\underline{\underline{y}}) - \hat{a}(\underline{\underline{y}})b(\underline{\underline{y}}) = 0$. We conclude that $a(\underline{\underline{y}})/b(\underline{\underline{y}}) = \hat{a}(\underline{\underline{y}})/\hat{b}(\underline{\underline{y}})$.

Examples: (1) Let $n = 3$, $k = \mathbb{Q}$, $\Omega = \mathbb{C}$, and V the sphere $X_1^2 + X_2^2 + X_3^2 - 1 = 0$. Let φ be the rational function represented by $1 = 1/1$. Put $\underline{y} = (1,0,0)$. Now φ is defined at $\underline{y}$ and $\varphi(\underline{y}) = 1$. Now φ is also represented by $1/(X_1^2 + X_2^2 + X_3^2)$. Again the denominator does not vanish at $\underline{y}$. If we use this representation, we again find, as expected, that $\varphi(\underline{y}) = 1$. Finally φ is also represented by $(X_1 - X_1^2 - X_2^2 - X_3^2)/(X_1 - 1)$. This representative cannot be used to compute $\varphi(\underline{y})$, since its denominator vanishes at $\underline{y}$.

(2) Let n, k, Ω and V be as above. Let φ be the rational function represented by $1/X_3$. This function φ is certainly defined if $\underline{y} \in V$ and $y_3 \neq 0$. We ask if there is representative of φ which allows us to define $\varphi(\underline{y})$ for some $\underline{y}$ with $y_3 = 0$. Let $a(\underline{X})/b(\underline{X})$ be a representative. Then

$$\frac{1}{X_3} - \frac{a(\underline{X})}{b(\underline{X})} = \frac{b(\underline{X}) - X_3 a(\underline{X})}{X_3 b(\underline{X})}$$

vanishes on V. Thus $b(\underline{X}) - a(\underline{X})\, X_3 \in (X_1^2 + X_2^2 + X_3^2 - 1)$. So $b(\underline{X}) \in (X_3, X_1^2 + X_2^2 + X_3^2 - 1)$, and therefore $b(\underline{y}) = 0$, if $\underline{y} \in V$ and $y_3 = 0$. It follows that φ is defined precisely for those points $\underline{y}$ on the sphere which are not on the circle $y_3 = 0$, $y_1^2 + y_2^2 - 1 = 0$.

THEOREM 3C. *Let φ be a rational function on a variety V. The set of points $\underline{y} \in V$ for which φ is not defined is a proper* algebraic subset of V .

Proof: The set of points where φ is not defined is

$$S = V \cap \bigcap_{b(\underline{\underline{X}})} A((b(\underline{\underline{X}})) \quad ,$$

where the intersection is taken over all $b(\underline{\underline{X}})$ which occur as a denominator of a representative of φ . Since the intersection of an arbitrary number of algebraic sets is an algebraic set, S is an algebraic set. In addition, S is a proper subset of V , since a generic point of V is not in S .

Let φ be a rational function of a variety V , and let W be a subvariety of V . We say φ is defined on W if φ is defined at a generic point of W .

A rational map $\underline{\underline{\varphi}}$ from a variety V to Ω^m is defined simply as an m-tuple of rational functions $(\varphi_1,\ldots,\varphi_m)$. We say $\underline{\underline{\varphi}}$ is defined at $\underline{\underline{y}} \in V$, if each $\varphi_i(\underline{\underline{y}})$ is defined at $\underline{\underline{y}}$. If this is the case, put $\underline{\underline{\varphi}}(\underline{\underline{y}}) = (\varphi_1(\underline{\underline{y}}),\ldots,\varphi_n(\underline{\underline{y}}))$. The set of points $\underline{\underline{y}} \in V$ for which $\underline{\underline{\varphi}}$ is not defined is the union of the sets of points for which φ_i is not defined $(i = 1,\ldots,m)$. In view of Theorem 3C , and since a finite union of proper algebraic subsets of a variety is still a proper algebraic subset, the points where $\underline{\underline{\varphi}}$ is not defined are a proper algebraic subset of V .

The image of $\underline{\underline{\varphi}}$ is defined as the closure of the set of points $\underline{\underline{\varphi}}(\underline{\underline{y}})$, $\underline{\underline{y}} \in V_A$ for which $\underline{\underline{\varphi}}$ is defined.

THEOREM 3D. The image of $\underline{\underline{\varphi}}$ is a variety W . If $\underline{\underline{x}}$ is a generic point of V , then $\underline{\underline{\varphi}}(\underline{\underline{x}})$ is a generic point of W .

Proof: Let $V = (\overline{\underline{x}})$. If $\underline{x} \to \underline{y}$ and if $\underline{\varphi}(\underline{y})$ is defined, we have to show that $\underline{\varphi}(\underline{x}) \to \underline{\varphi}(\underline{y})$. Let $\underline{\varphi} = (\varphi_1,\dots,\varphi_m)$, and suppose that φ_i is represented by $a_i(\underline{X})/b_i(\underline{X})$ with $b_i(\underline{y}) \neq 0$. Let $f(\underline{\varphi}(\underline{x})) = 0$, and suppose that $f(\underline{U}) = f(U_1,\dots,U_m)$ is of degree d_i in U_i. Put

$$g(U_1,\dots,U_m, V_1,\dots,V_m) = V_1^{d_1}\dots V_m^{d_m} f\left(\frac{U_1}{V_1},\dots,\frac{U_m}{V_m}\right) .$$

Since $f(a_1(\underline{x})/b_1(\underline{x}),\dots,a_m(\underline{x})/b_m(\underline{x})) = 0$, it follows that $g(a_1(\underline{x}),\dots,a_m(\underline{x}),b_1(\underline{x}),\dots,b_m(\underline{x})) = 0$. But $\underline{x} \to \underline{y}$, so $g(a_1(\underline{y}),\dots,a_m(\underline{y}),b_1(\underline{y}),\dots,b_m(\underline{y})) = 0$, and

$$b_1(\underline{y})^{d_1}\dots b_m(\underline{y})^{d_m} f\left(\frac{a_1(\underline{y})}{b_1(\underline{y})},\dots,\frac{a_m(\underline{y})}{b_m(\underline{y})}\right) = 0 .$$

Since $b_1(\underline{y})^{d_1}\dots b_m(\underline{y})^{d_m} \neq 0$, it follows that

$$f(\underline{\varphi}(\underline{y})) = f\left(\frac{a_1(\underline{y})}{b_1(\underline{y})},\dots,\frac{a_m(\underline{y})}{b_m(\underline{y})}\right) = 0 .$$

So every polynomial f vanishing on $\underline{\varphi}(\underline{x})$ also vanishes on $\underline{\varphi}(\underline{y})$, and $\underline{\varphi}(\underline{x}) \to \underline{\varphi}(\underline{y})$.

Example: Let V be the sphere $x_1^2 + x_2^2 + x_3^2 = 1$, and let $\underline{\varphi}: V \to \Omega^2$ have a representation as $\underline{\varphi} = ((X_1^2 + X_2^2)/X_3^2 , -1/X_3^2)$. Let $\underline{\xi} = (\xi_1,\xi_2,\xi_3)$ be a generic point of V. We have

$$\underline{\varphi}(\underline{\xi}) = \left(\frac{\xi_1^2 + \xi_2^2}{\xi_3^2} , -\frac{1}{\xi_3^2}\right) = \left(\frac{1}{\xi_3^2} - 1 , -\frac{1}{\xi_3^2}\right).$$

Thus $\underline{\underline{\varphi}}(\underline{\underline{\xi}}) = (\zeta_1,\zeta_2)$ satisfies $\zeta_1 + \zeta_2 + 1 = 0$. Since $\underline{\underline{\varphi}}(\underline{\underline{\xi}})$ has transcendence degree 1, it is in fact a generic point of the line $z_1 + z_2 + 1 = 0$. Thus this line is the image of $\underline{\underline{\varphi}}$. But not every point on this line is of the type $\underline{\underline{\varphi}}(\underline{\underline{y}})$. If (z_1,z_2) is on the line and is $\neq (-1,0)$, then if we pick y_1,y_2,y_3 in Ω with $y_3 = 1/\sqrt{z_2}$, $y_1^2 + y_2^2 + y_3^2 - 1 = 0$, we obtain $\underline{\underline{\varphi}}(\underline{\underline{y}}) = (z_1,z_2)$. But $(z_1,z_2) = (-1,0)$ is not of the type $\underline{\underline{\varphi}}(\underline{\underline{y}})$. For if $y_3 \neq 0$, then $\underline{\underline{\varphi}}(\underline{\underline{y}}) \neq (-1,0)$, and if $y_3 = 0$, then $\underline{\underline{\varphi}}(\underline{\underline{y}})$ is not defined.

THEOREM 3E. *Let $\underline{\underline{\varphi}}$ be a rational map from V with image W. Let T be a proper algebraic subset of W. Then the set $L \subseteq V$ consisting of points $\underline{\underline{y}}$ where either $\underline{\underline{\varphi}}$ is not defined or where $\underline{\underline{\varphi}}(\underline{\underline{y}}) \in T$, is a proper algebraic subset of V.*

Proof: Suppose W and T lie in Ω^m. Suppose T is defined by equations $g_1(\underline{\underline{y}}) = \dots = g_t(\underline{\underline{y}}) = 0$, where $\underline{\underline{y}} = (y_1,\dots,y_m)$. Let $g_i(Y_1,\dots,Y_m)$ have degree d_{ij} in Y_j $(1 \leq i \leq t,\ 1 \leq j \leq m)$. Put

$$h_i(Y_1,\dots,Y_m, Z_1,\dots,Z_m) = Z^{d_{i1}} \dots Z^{d_{im}} g_i\left(\frac{Y_1}{Z_1},\dots,\frac{Y_m}{Z_m}\right).$$

Let

$$\underline{\underline{r}} = \underline{\underline{r}}(\underline{\underline{X}}) = (a_1(\underline{\underline{X}})/b_1(\underline{\underline{X}}),\dots,a_m(\underline{\underline{X}})/b_m(\underline{\underline{X}}))$$

represent $\underline{\underline{\varphi}}$ and put

$$\ell_i^{\underline{\underline{r}}}(\underline{\underline{X}}) = b_1(\underline{\underline{X}})\dots b_m(\underline{\underline{X}})\ h_i(a_1(\underline{\underline{X}}),\dots,a_m(\underline{\underline{X}}),b_1(\underline{\underline{X}}),\dots,b_m(\underline{\underline{X}}))\quad (1 \leq i \leq t).$$

Let $L_{\underline{\underline{r}}}$ consist of points $\underline{\underline{y}}$ of V with

$$\ell_1^{\underline{\underline{r}}}(\underline{\underline{y}}) = \dots = \ell_t^{\underline{\underline{r}}}(\underline{\underline{y}}) = 0.$$

We claim that

(3.1) $$L = \bigcap L_{\underline{\underline{r}}} ,$$

with the intersection taken over all representations $\underline{\underline{r}}$ of $\underline{\varphi}$. In fact if $\underline{\underline{y}} \notin L_{\underline{\underline{r}}}$ for some $\underline{\underline{r}}$, then some $\ell_i^{\underline{\underline{r}}}(\underline{y}) \neq 0$, and hence $b_1(\underline{\underline{y}}_m) \ldots b_m(\underline{y}) \neq 0$ and $g_i(a_1(\underline{y})/b_1(\underline{y}), \ldots, a_m(\underline{y})/b_m(\underline{y})) \neq 0$. So $\underline{\varphi}(\underline{y})$ is defined and $g_i(\underline{\varphi}(\underline{y})) \neq 0$, so that $\underline{\varphi}(\underline{y}) \notin T$ and $\underline{y} \notin L$. On the other hand if $\underline{y} \notin L$, then $\underline{\varphi}(\underline{y})$ is defined, and for some representation $\underline{\underline{r}}$ we have $b_1(\underline{y}) \ldots b_m(\underline{y}) \neq 0$. Moreover, $\underline{\varphi}(\underline{y}) \notin T$, whence some $g_i(\underline{\varphi}(\underline{y})) \neq 0$, and $\ell_i^{\underline{\underline{r}}}(\underline{y}) \neq 0$. Thus $\underline{y} \notin L_{\underline{\underline{r}}}$, and (3.1) is established.

In view of (3.1), L is an algebraic subset of V. Since a generic point of V lies outside each $L_{\underline{\underline{r}}}$, the set L is a proper algebraic subset.

<u>Example</u>. Let $V \subseteq \Omega^3$ be the sphere $x_1^2 + x_2^2 + x_3^2 - 1 = 0$ and let $W \subseteq \Omega^2$ be the line $z_1 + z_2 + 1 = 0$. We have seen above that the map $\underline{\varphi}$ represented by $((X_1^2 + X_2^2)/X_3^2 , -1/X_3^2)$ has image W. Let $T \subseteq W$ consist of the single point $(0,-1)$. It is easily seen that the set L of points $\underline{y}$ where $\underline{\varphi}(\underline{y})$ is not defined or where $\underline{\varphi}(\underline{y}) \in T$ consists of $\underline{y} \in V$ with $y_3(y_3^2 - 1) = 0$.

4. Birational Maps.

We define a rational map from a variety V to a variety W as a rational map $\underline{\underline{\varphi}}$ of V whose image is contained in W. We express this in symbols by $\underline{\underline{\varphi}}: V \rightarrow W$.

Let $\underline{\underline{\varphi}}: V \rightarrow W$ and $\underline{\underline{\psi}}: W \rightarrow U$ be rational maps such that $\underline{\underline{\psi}}$ is defined on the image of V under $\underline{\underline{\varphi}}$. Thus if $\underline{\underline{x}}$ is a generic point of V, then $\underline{\underline{\psi}}$ is defined on $\underline{\underline{\varphi}}(\underline{\underline{x}})$. Suppose $V \subseteq \Omega^v$, $W \subseteq \Omega^w$, $U \subseteq \Omega^u$, and suppose $\underline{\underline{\varphi}}$ is represented by

$$(4.1) \qquad (a_1(\underline{\underline{X}})/b_1(\underline{\underline{X}}),\ldots,a_w(\underline{\underline{X}})/b_w(\underline{\underline{X}})) ,$$

and $\underline{\underline{\psi}}$ is represented by

$$(4.2) \qquad (c_1(\underline{\underline{Y}})/d_1(\underline{\underline{Y}}),\ldots,c_u(\underline{\underline{Y}})/d_u(\underline{\underline{Y}})) ,$$

where $d_1,\ldots,d_u$ are non-zero at $\underline{\underline{\varphi}}(\underline{\underline{x}})$. Let $\underline{\underline{\psi}}\,\underline{\underline{\varphi}}$ be the rational map $V \rightarrow U$ represented by

$$(4.3) \qquad (c_1(a_1(\underline{\underline{X}})/b_1(\underline{\underline{X}}),\ldots)/d_1(a_1(\underline{\underline{X}})/b_1(\underline{\underline{X}}),\ldots)),\ldots,c_u(\ldots)/d_u(\ldots)) .$$

Since $d_1,\ldots,d_u$ are not zero at $\underline{\underline{\varphi}}(\underline{\underline{x}})$, each of the u components in (4.3) lies in $\mathcal{O}_{\underline{\underline{x}}}$, and $\underline{\underline{\psi}}\,\underline{\underline{\varphi}}(\underline{\underline{x}})$ is defined and equals $\underline{\underline{\psi}}(\underline{\underline{\varphi}}(\underline{\underline{x}}))$. It is clear that $\underline{\underline{\psi}}\,\underline{\underline{\varphi}}$ is independent of the special representations (4.1), (4.2) of $\underline{\underline{\varphi}}$, $\underline{\underline{\psi}}$, respectively. We call $\underline{\underline{\psi}}\,\underline{\underline{\varphi}}$ the <u>composite</u> of $\underline{\underline{\psi}}$ and $\underline{\underline{\varphi}}$. If $\underline{\underline{v}}$ is a point of V such that $\underline{\underline{\varphi}}$ is defined at $\underline{\underline{v}}$ and $\underline{\underline{\psi}}$ is defined at $\underline{\underline{\varphi}}(\underline{\underline{v}})$, then $\underline{\underline{\psi}}\,\underline{\underline{\varphi}}$ is defined at $\underline{\underline{v}}$ and

$$\underline{\underline{\psi}}\,\underline{\underline{\varphi}}(\underline{\underline{v}}) = \underline{\underline{\psi}}(\underline{\underline{\varphi}}(\underline{\underline{v}})) .$$

But $\underline{\underline{\psi}}\underline{\underline{\varphi}}(\underline{\underline{v}})$ may be defined although perhaps either $\underline{\underline{\varphi}}(\underline{\underline{v}})$ is not defined, or $\varphi(\mathrm{v})$ is defined and $\psi(\varphi(\mathrm{v}))$ is not defined.

Examples. (1) Let $V = \Omega^1$, $W = \Omega^2$, $U = V = \Omega^1$. Further let $\varphi\colon V \to W$ be represented by (X^2, X), and let $\psi\colon W \to V$ be represented by X_1/X_2. Then $\psi\varphi$ is the identity map on V. Thus $\psi\varphi$ is defined on 0 and $\psi\varphi(0) = 0$. However $\varphi(0) = (0,0)$, and ψ is not defined at $(0,0)$.

(2) Let $k = \mathbb{Q}$ and $\Omega = \mathbb{C}$. Let $V = \Omega^1$, W the unit circle $x_1^2 + x_2^2 - 1 = 0$, and $U = V = \Omega^1$. Further let $\varphi\colon V \to W$ be represented by $(2X/(X^2+1), (X^2-1)/(X^2+1))$, and let $\psi\colon W \to V$ be represented by $X_1/(1 - X_2)$. Then $\psi\varphi$ is the identity map on V and $\varphi\psi$ is the identity map on W. In particular, $\psi\varphi$ is defined at i and $\psi\varphi(i) = i$, but φ is not defined at i.

Exercise. Show that in Example (2), φ is defined for every point of V except for i, $-i$, and that ψ is defined for every point of W except for $(0,1)$. Further show that every point of V with the exception of $i, -i$ is of the type $\psi(\mathrm{y})$ with $\mathrm{y} \in W$, and every point of W with the exception of $(0,1)$ is of the type $\varphi(\mathrm{x})$ with $\mathrm{x} \in V$. Hence if V' is obtained from V by deleting i, $-i$ and W' is obtained from W by deleting $(0,1)$, then φ and ψ provide a 1-1 correspondence between points of V' and of W'.

A rational map $\varphi\colon V \to W$ is called a *bi-rational map* (or a *bi-rational correspondence*) if there exists a rational map $\psi\colon W \to V$ such that $\psi\varphi$ is the identity on V and $\varphi\psi$ is the identity on W. Two varieties are *bi-rationally equivalent* if there exists a bi-rational correspondence between them. We denote this by $V \cong W$. This is an equivalence relation of varieties. (Note that this relation is defined in terms of the ground field k).

THEOREM 4A. *Let* $\underline{\underline{\varphi}}$ *be a bi-rational map from* V *to* W *with inverse* $\underline{\underline{\psi}}$. *Then there exist proper algebraic subsets* L *of* V *and* M *of* W, *such that on the set theoretic differences* $V \sim L$ *and* $W \sim M$, *the maps* $\underline{\underline{\varphi}}$ *and* $\underline{\underline{\psi}}$ *are defined everywhere and are inverses of each other.*

Proof: Let S be the subset of V where $\underline{\underline{\varphi}}$ is not defined. Let T be the subset of W where $\underline{\underline{\psi}}$ is not defined. Let L be the subset of V where either $\underline{\underline{\varphi}}$ is not defined or where $\underline{\underline{\varphi}}(\underline{\underline{x}}) \in T$. Similarly, let M be the subset of W where either $\underline{\underline{\psi}}$ is not defined or where $\underline{\underline{\psi}}(\underline{\underline{x}}) \in S$. In view of Theorem 3E, the sets L, M are proper algebraic subsets of V, W, respectively. Now $\underline{\underline{\varphi}}$ is defined on $V \sim L$. Clearly, if $\underline{\underline{x}} \in V \sim L$, then $\underline{\underline{\varphi}}(\underline{\underline{x}}) \notin T$. So $\underline{\underline{\psi}}(\underline{\underline{\varphi}}(\underline{\underline{x}}))$ is defined; but then $\underline{\underline{\psi}}(\underline{\underline{\varphi}}(\underline{\underline{x}})) = \underline{\underline{x}}$. From this it follows that $\underline{\underline{\varphi}}(\underline{\underline{x}}) \in W \sim M$, since $\underline{\underline{x}} \notin S$. So the restriction of $\underline{\underline{\varphi}}$ to $V \sim L$ maps $V \sim L$ into $W \sim M$. The restriction of $\underline{\underline{\psi}}$ to $W \sim M$ maps $W \sim M$ into $V \sim L$. These maps are inverses of each other.

THEOREM 4B. *Let* V *and* W *be varieties. Then* $V \cong W$ *if and only if their function fields are* k*-isomorphic.*

Proof: If $\underline{\underline{x}}$ is a generic point of V and $\underline{\underline{y}}$ is a generic point of W, then the function fields are isomorphic to $k(\underline{\underline{x}})$ and $k(\underline{\underline{y}})$, respectively. So we need to show that $V \cong W$ if and only if $k(\underline{\underline{x}})$ is isomorphic to $k(\underline{\underline{y}})$.

Suppose that $V \cong W$. Let $\underline{\underline{\varphi}}: V \to W$ and $\underline{\underline{\psi}}: W \to V$ be bi-rational maps, such that $\underline{\underline{\varphi}}\,\underline{\underline{\psi}}$ and $\underline{\underline{\psi}}\,\underline{\underline{\varphi}}$ are the identity maps on W and V, respectively.

It is clear from Theorem 4A that the "image" of V under $\underline{\underline{\varphi}}$ is W. Thus if $\underline{\underline{x}}$ is a generic point of V, then by Theorem 3D the point $\underline{\underline{y}} = \underline{\underline{\varphi}}(\underline{\underline{x}})$ is a generic point of W. We have $\underline{\underline{y}} = \underline{\underline{\varphi}}(\underline{\underline{x}})$ and $\underline{\underline{x}} = \underline{\underline{\psi}}(\underline{\underline{y}})$, whence $k(\underline{\underline{y}}) \subseteq k(\underline{\underline{y}})$ and $k(\underline{\underline{x}}) \subseteq k(\underline{\underline{y}})$, whence $k(\underline{\underline{x}}) = k(\underline{\underline{y}})$. Thus the function fields are certainly k-isomorphic.

Conversely, let $k(\underline{\underline{x}})$ be isomorphic to $k(\underline{\underline{y}})$, where $\underline{\underline{x}} = (x_1 \ldots, x_n), \underline{\underline{y}} = (y_1, \ldots, y_m)$ are generic points of V, W respectively. Let α be a k-isomorphism from $k(\underline{\underline{x}})$ to $k(\underline{\underline{y}})$. Let $\alpha(x_i) = x_i'$ $(i = 1, \ldots, n)$ and put $\underline{\underline{x}}' = (x_1', \ldots, x_n')$. Then $k(\underline{\underline{x}}') = k(\underline{\underline{y}})$ and $\underline{\underline{x}}'$ is again a generic point of V. Thus we may suppose that $k(\underline{\underline{x}}) = k(\underline{\underline{y}})$. Suppose that

$$y_i = r_i(\underline{\underline{x}}) \qquad (i = 1, \ldots, m)$$

and

$$x_j = s_j(\underline{\underline{y}}) \qquad (j = 1, \ldots, n)$$

for certain rational functions $r_1, \ldots, r_m$ and $s_1, \ldots, s_n$. Then $\underline{\underline{\varphi}}: V \to W$ represented by $(r_1(\underline{\underline{X}}), \ldots, r_m(\underline{\underline{X}}))$ and $\underline{\underline{\psi}}: W \to V$ represented by $(s_1(\underline{\underline{Y}}), \ldots, s_n(\underline{\underline{Y}}))$ are rational maps which are inverses of each other.

In §3 we defined a rational curve as one whose function field is isomorphic to $k(X)$. In view of Theorem 4B, we may also define a rational curve as a curve which is birationally equivalent to Ω^1.

<u>LEMMA 4C</u>. <u>The following two conditions on a field</u> k <u>are equivalent</u>.

(i) <u>Either char</u> $k = 0$, <u>or char</u> $k = p > 0$ <u>and for every</u> $a \in k$ <u>there is a</u> $b \in k$ <u>with</u> $b^p = a$.

(ii). Every algebraic extension of k is separable.

Proof. We clearly may suppose that char $k = p > 0$.

(i) → (ii). A polynomial of $k[X]$ of the type

$$a_0 + a_1 X^p + \dots + a_t X^{tp} \tag{4.4}$$

equals $(b_0 + b_1 X + \dots + b_t X^t)^p$ where $b_i^p = a_i$ $(i = 0, \dots, t)$. Thus an irreducible polynomial over k is not of the type (4.4), hence is separable.

(ii) → (i). Suppose there is an $a \in k$ not of the type $a = b^p$ with $b \in k$. Then there is a b which is not in k but in an algebraic extension of k, with $a = b^p$. Since p is a prime, it is easily seen that $i = p$ is the smallest positive exponent with $b^i \in k$. The polynomial $X^p - a = (X - b)^p$ has proper factors $(X - b)^i$ with $1 \leqq i \leqq p - 1$, but none of these factors lies in $k[X]$ since $b^i \notin k$. Thus $X^p - a$ is irreducible over k , and b is inseparable over k .

A field with the properties of the lemma is called perfect. A Galois field is perfect. For if a lies in the finite field F_q with $q = p^\nu$ elements, then $a = a^q = \left(a^{p^{\nu-1}}\right)^p$.

THEOREM 4D. Suppose V is a variety defined over a perfect ground field k . Then V is birationally equivalent to a hypersurface.

Proof. Suppose dim $V = d$ and $\underline{\underline{x}} = (x_1, \dots, x_n)$ is a generic point of V . Then $n \geqq d$. In view of Theorem 4B it will suffice to show that there is a $\underline{\underline{y}} = (y_1, \dots, y_{d+1})$ with

(4.5) $$k(\underline{\underline{x}}) = k(\underline{\underline{y}}) .$$

We shall show this by induction on $n-d$. If $n-d=0$, set $y_1 = x_1, \ldots, y_d = x_d$, $y_{d+1} = 0$. If $n-d=1$, set $\underline{\underline{y}} = \underline{\underline{x}}$. Suppose now that $n-d>1$ and that our claim is true for smaller values of $n-d$. We may suppose without loss of generality that $x_1, \ldots, x_{d+1}$ have transcendence degree d over k. Then $(x_1, \ldots, x_{d+1})$ is the generic point of a hypersurface in Ω^{d+1}. This hypersurface is defined by an equation $f(z_1, \ldots, z_{d+1}) = 0$ where $f(Z_1, \ldots, Z_{d+1})$ is irreducible over k. Since k is perfect, it is clear that f is not a polynomial in $Z_1^p, \ldots, Z_{d+1}^p$ if char $k = p > 0$. We may then suppose without loss of generality that f is not a polynomial in $Z_1, \ldots, Z_d, Z_{d+1}^p$. Thus f is separable in the variable Z_{d+1}, and x_{d+1} is separable algebraic over $k(x_1, \ldots, x_d)$. By the theorem of the primitive element (see Van der Waerden, §43), there is an x' with

$$k(x_1, \ldots, x_d, x_{d+1}, x_{d+2}) = k(x_1, \ldots, x_d, x').$$

Thus $\underline{\underline{x}}' = (x_1, \ldots, x_d, x', x_{d+3}, \ldots, x_n)$ has $k(\underline{\underline{x}}') = k(\underline{\underline{x}})$. By induction hypothesis there is a $\underline{\underline{y}} \in \Omega^{d+1}$ with $k(\underline{\underline{x}}') = k(\underline{\underline{y}})$, hence with (4.5).

5. Linear Disjointness of Fields

LEMMA 5A: *Suppose that* Ω , K , L , k *are fields with* $k \subseteq K \subseteq \Omega$, $k \subseteq L \subseteq \Omega$:

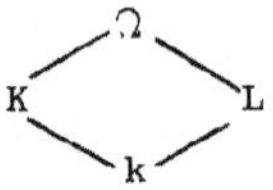

The following two properties are equivalent:

(i) *If elements* $x_1,\dots,x_m$ *of* K *are linearly independent over* k , *then they are also linearly independent over* L .

(ii) *If elements* $y_1,\dots,y_n$ *of* L *are linearly independent over* k , *then they are also linearly independent over* K .

Proof: By symmetry it is sufficient to show that (i) implies (ii). Let $y_1,\dots,y_n$ of L be linearly independent over k . Let $x_1,\dots,x_n$ of K be not all zero. We want to show that

$$x_1y_1 + \dots + x_ny_n \neq 0 . \tag{5.1}$$

Let d be the maximum number of $x_1,\dots,x_n$ which are linearly independent over k . Without loss of generality, we may assume that $x_1,\dots,x_d$ are linearly independent over k . Thus for $d < i \leqq n$ we have $x_i = \sum_{j=1}^{d} c_{ij}\, x_j$, where $c_{ij} \in k$. We obtain

$$x_1y_1 + \dots + x_ny_n = \left(y_1 + \sum_{i=d+1}^{n} c_{i1}\, y_i\right) x_1 + \dots$$

$$+ \left(y_d + \sum_{i=d+1}^{n} c_{id}\, y_i\right) x_d \quad .$$

Here $x_1,\ldots,x_d \in K$ are linearly independent over k, whence linearly independent over K. Their coefficients are not zero since $y_1,\ldots,$ y_n are linearly independent over k. Thus (5.1) follows.

We say that field extensions K, L of k are <u>linearly disjoint over k</u>, if properties (i) and (ii) hold.

<u>Examples:</u> (i) Consider the fields

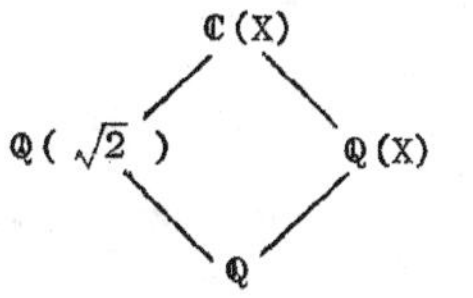

Here $\mathbb{Q}(\sqrt{2})$ and $\mathbb{Q}(X)$ are linearly disjoint over $\mathbb{Q}$. For if $(a + b\sqrt{2}$) and $c + d\sqrt{2}$) are linearly independent over $\mathbb{Q}$, then clearly they are linearly independent over $\mathbb{Q}(X)$.

(ii) Let X,Y,Z,W be variables, and consider the fields

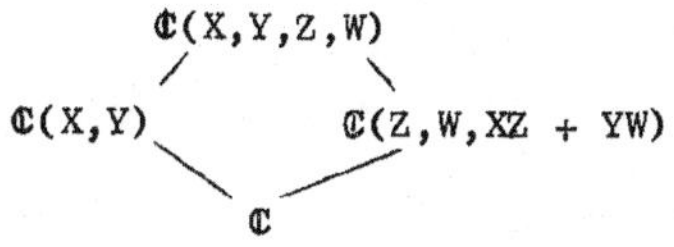

In this case $\mathbb{C}(X,Y)$ and $\mathbb{C}(Z,W,XZ + YW)$ are not linearly disjoint over $\mathbb{C}$. For $Z,W,XZ + YW$ are linearly dependent over $\mathbb{C}(X,Y)$, but are linearly independent over $\mathbb{C}$.

LEMMA 5B: Let us consider fields

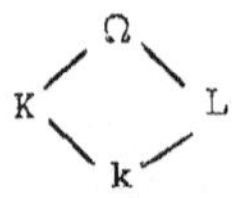

,

where L *is the quotient field of a ring* R . *For linear disjointness it is sufficient to show that if* $z_1,\ldots,z_n \in R$ *are linearly independent over* k , *then they are also linearly independent over* K .

Proof: Let $y_1,\ldots,y_n \in L$ be linearly independent over k . We can find a $z \neq 0$, $z \in R$, such that $zy_1,\ldots,zy_n \in R$. Now $zy_1,\ldots,zy_n$ are linearly independent over k, hence also linearly independent over K . Therefore $y_1,\ldots,y_n$ are linearly independent over K .

LEMMA 5C: Suppose we have fields

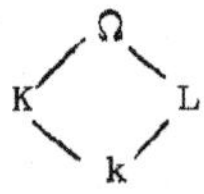

,

where K *is algebraic over* k . *Let* KL *be the set of expressions* $x_1 y_1 + \ldots + x_n y_n$ *with* $x_i \in K$, $y_i \in L$ *for* $1 \leq i \leq n$, *and with* n *arbitrary*.

(i) *The set* KL *is a field, it contains* K *and* L , *and is the smallest such field.*

(ii) *Suppose that* $[K : k]$ *is finite. Then* $[KL : L] \leqq [K : k]$, *with equality precisely if* K , L *are linearly disjoint over* k .

(iii) Now suppose that K, L are linearly disjoint over k. Let α be a k-isomorphism from K to a field H containing k. Let β be a k-isomorphism from L to H. Then

$$x_1 y_1 + \dots + x_n y_n \to \alpha(x_1)\,\beta(y_1) + \dots + \alpha(x_n)\,\beta(y_n)$$

is a well-defined map from KL to H. It is a k-isomorphism into H.

Proof: Exercise.

LEMMA 5D. Suppose we have a diagram of fields and subfields

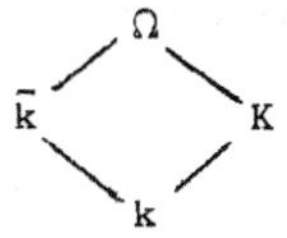

where k is perfect and $\bar{k}$ is the algebraic closure of k. Then K, $\bar{k}$ are linearly disjoint over k if and only if k is algebraically closed in K.

Proof: If k is not algebraically closed in K, then there exists a proper algebraic extension k_1 of k with $k_1 \subseteq K$;

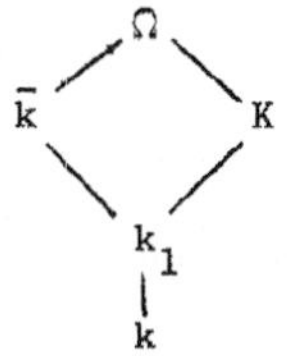

It is now clear that $\bar{k}$ and K cannot be linearly disjoint over k.

Conversely, suppose that k is algebraically closed in K. It suffices to show that k_2, K are linearly disjoint over k, where k_2 is any finite algebraic extension of k. Since k is perfect, $k_2 = k(x)$, and we have the following diagram of fields:

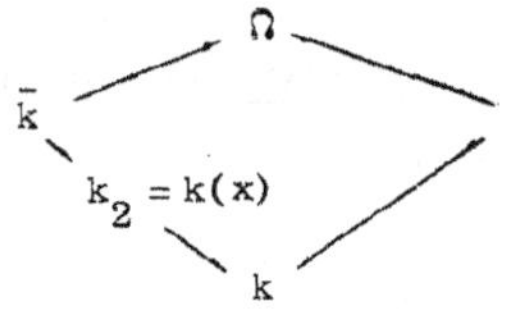

If $f(X)$ is the defining polynomial of x over k, then it remains irreducible over K, since every proper factor of $f(X)$ has coefficients which are algebraic over k, with some coefficients not in k, and hence not in K.

So for the fields

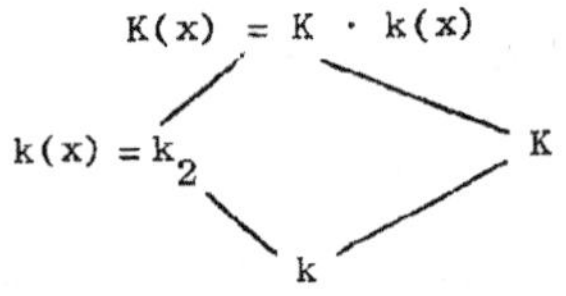

we have $[K\cdot k(x) : K] = [k(x) : k]$; hence $k(x), K$ are linearly disjoint over k by Lemma 5C.

6. Constant Field Extensions

Consider fields k, K, Ω, such that $k \subsetneq K \subseteq \Omega$, and Ω is algebraically closed and has infinite transcendence degree over K. If $\underline{\underline{x}} \in \Omega^n$, then $\mathfrak{J}_k^{\dagger}(\underline{\underline{x}})$ is the ideal of all polynomials $f(\underline{\underline{X}}) \in k[\underline{\underline{X}}]$ with $f(\underline{\underline{x}}) = 0$. We have seen in §1 that $\mathfrak{J}_k(\underline{\underline{x}}) = \mathcal{Y}$ is a

† Given a subset $M \subseteq \Omega^n$, we write $\mathfrak{J}_k(M)$ or $\mathfrak{J}_K(M)$ for the set of polynomials $f(\underline{\underline{X}})$ in $k[\underline{\underline{X}}]$ or $K[\underline{\underline{X}}]$, respectively, which vanish on M.

prime ideal in $k[\underline{\underline{X}}]$. Similarly, $\mathfrak{I}_K(\underline{\underline{X}}) = \mathfrak{P}$ is a prime ideal in $K[\underline{\underline{X}}]$. Let $\mathfrak{y}K[\underline{\underline{X}}]$ be the ideal in $K[\underline{\underline{X}}]$ generated by $\mathfrak{y}$. The ideal $\mathfrak{y}K[\underline{\underline{X}}]$ consists of all linear combinations $c_1 f_1 + \dots + c_m f_m$, where $c_i \in K$, $f_i \in \mathfrak{y}$ $(i = 1,\dots,m)$. Clearly $\mathfrak{y}K[\underline{\underline{X}}] \subseteq \mathfrak{P}$. Denote the closure of a point $\underline{\underline{x}}$ with respect to k, K by $(\overline{\underline{\underline{x}}})^k$, $(\overline{\underline{\underline{x}}})^K$, respectively. We have $(\overline{\underline{\underline{x}}})^k = A(\mathfrak{y}) = A(\mathfrak{y}K[\underline{\underline{X}}]) \supseteq A(\mathfrak{M}) = (\overline{\underline{\underline{x}}})^K$. So

$$(\overline{\underline{\underline{x}}})^K \subseteq (\overline{\underline{\underline{x}}})^k .$$

<u>Example:</u> Let $k = \mathbb{Q}$, $K = \mathbb{Q}(\sqrt{2})$, $\Omega = \mathbb{C}$, and $n = 2$. Consider the point $(e\sqrt{2}, e) = \underline{\underline{x}}$. Then $(\overline{\underline{\underline{x}}})^k$ is the set of zeros of the polynomial $X^2 - 2Y^2$. But $(\overline{\underline{\underline{x}}})^K$ is the set of zeros of $X - \sqrt{2}\,Y$.

<u>THEOREM 6A. Let</u> $k \subseteq K \subseteq \Omega$ <u>be fields, where</u> Ω <u>is algebraically closed and has infinite transcendence degree over</u> K. <u>Let</u> $\underline{\underline{x}} \in \Omega^n$, $\mathfrak{I}_k(\underline{\underline{x}}) = \mathfrak{y}$, $\mathfrak{I}_K(\underline{\underline{x}}) = \mathfrak{M}$. <u>Consider the following four properties:</u>

(i) <u>The fields</u> K, $k(\underline{\underline{x}})$ <u>are linearly disjoint extensions of</u> k,

(ii) $\mathfrak{M} = \mathfrak{y}K[\underline{\underline{X}}]$,

(iii) $(\overline{\underline{\underline{x}}})^k = (\overline{\underline{\underline{x}}})^K$,

(iv) $\mathfrak{P} = \sqrt{\mathfrak{y}K[\underline{\underline{X}}]}$.

<u>The properties</u> (i), (ii) <u>are equivalent. Property</u> (ii) <u>implies property</u> (iii), <u>which in turn implies property</u> (iv).

<u>Proof:</u> To show that (i) implies (ii), let $f(\underline{\underline{X}}) \in \mathfrak{M}$. Write $f(\underline{\underline{X}}) = \sum_{i=1}^{n} a_i f_i(\underline{\underline{X}})$, where $a_i \in K$, $f_i(\underline{\underline{X}}) \in k[\underline{\underline{X}}]$, and $a_1,\dots,a_n$ are linearly independent over k. Now $f(\underline{\underline{x}}) = 0$, so $\sum_{i=1}^{n} a_i f_i(\underline{\underline{x}}) = 0$.

By the linear disjointness of K and $k(\underline{\underline{x}})$, the a_i's are linearly independent over $k(\underline{\underline{x}})$. It follows that each $f_i(\underline{\underline{x}}) = 0$, and each $f_i(\underline{\underline{X}}) \in \mathcal{Y}$. Thus $f(\underline{\underline{X}}) \in \mathcal{Y} K[\underline{\underline{X}}]$.

To show that (ii) implies (i), let $u_1(\underline{\underline{X}}),\ldots,u_\ell(\underline{\underline{X}})$ be elements of $k[\underline{\underline{X}}]$, such that $u_1(\underline{\underline{x}}),\ldots,u_\ell(\underline{\underline{x}})$ are linearly independent over k. By Lemma 5B, it will suffice to show that $u_1(\underline{\underline{x}}),\ldots,u_\ell(\underline{\underline{x}})$ remain linearly independent over K. Suppose $a_1u_1(\underline{\underline{x}}) + \cdots + a_\ell u_\ell(\underline{\underline{x}}) = 0$, with $a_i \in K$. Let $f(\underline{\underline{X}}) = a_1u_1(\underline{\underline{X}}) + \cdots + a_\ell u_\ell(\underline{\underline{X}})$. Since $f(\underline{\underline{x}}) = 0$, the polynomial $f(\underline{\underline{X}})$ lies in $\mathfrak{M} = \mathcal{Y} K[\underline{\underline{X}}]$. We have a relation

$$(6.1) \qquad a_1u_1(\underline{\underline{X}}) + \cdots + a_\ell u_\ell(\underline{\underline{X}}) = b_1f_1(\underline{\underline{X}}) + \cdots + b_mf_m(\underline{\underline{X}}) ,$$

where $b_i \in K$, $f_i(\underline{\underline{X}}) \in \mathcal{Y}$ $(i = 1,\ldots,m)$, We may assume that $f_1,\ldots,f_m$ are linearly independent over k. We <u>claim that $u_1(\underline{\underline{X}}),\ldots,u_\ell(\underline{\underline{X}})$, $f_1(\underline{\underline{X}}),\ldots,f_m(\underline{\underline{X}})$ are linearly independent over</u> k. Suppose that

$$(6.2) \qquad \sum_{i=1}^{\ell} c_iu_i(\underline{\underline{X}}) + \sum_{j=1}^{m} d_jf_j(\underline{\underline{X}}) = 0 ,$$

where c_i, $d_j \in k$. Substituting $\underline{\underline{x}}$ for $\underline{\underline{X}}$, we obtain $\sum_{i=1}^{\ell} c_iu_i(\underline{\underline{x}}) = 0$. However, the $u_i(\underline{\underline{x}})$ are linearly independent over k, so that $c_1,\ldots,c_\ell$ are all zero. Thus (6.2) reduces to $\sum_{j=1}^{m} d_jf_j(\underline{\underline{X}}) = 0$. But the $f_j(\underline{\underline{X}})$ are linearly independent over k, and hence $d_1 = \cdots = d_m = 0$. We have established the linear independence of $u_1(\underline{\underline{X}}),\ldots,u_\ell(\underline{\underline{X}})$, $f_1(\underline{\underline{X}}),\ldots,f_m(\underline{\underline{X}})$ over k. These $\ell + m$ polynomials have coefficients in k and are linearly independent over k, and hence they are also linearly independent over

K^t. Hence in (6.1), all the coefficients are zero, and in particular $a_1 = \ldots = a_\ell = 0$.

We next want to show that (ii) implies (iii). Let $\underline{\underline{y}} \in (\overline{\underline{\underline{x}}})^k$. Then $f(\underline{\underline{y}}) = 0$ if $f(\underline{\underline{X}}) \in \mathfrak{y}$. Since $\mathfrak{P} = \mathfrak{y}\, k[\underline{\underline{X}}]$, we have $g(\underline{\underline{y}}) = 0$ for every $g(\underline{\underline{X}}) \in \mathfrak{P}$. Thus $\underline{\underline{y}} \in A(\mathfrak{P}) = (\overline{\underline{\underline{x}}})^K$. Hence $(\overline{\underline{\underline{x}}})^k \subseteq (\overline{\underline{\underline{x}}})^K$, and since the reversed relation is always true, we obtain (iii).

Finally, we are going to show that (iii) implies (iv). Suppose $f(\underline{\underline{X}}) \in \mathfrak{P}$. Then f vanishes on $(\overline{\underline{\underline{x}}})^K = (\overline{\underline{\underline{x}}})^k$, and $f \in \mathfrak{I}_K(\underline{\underline{x}}) = \mathfrak{I}_K((\overline{\underline{\underline{x}}})^k) = \mathfrak{I}_K(A(\mathfrak{y}K[\underline{\underline{X}}])) = \sqrt{\mathfrak{y}\, K[\underline{\underline{X}}]}$. So $\mathfrak{P} \subseteq \sqrt{\mathfrak{y}\, K[X]}$. Conversely, we have $\mathfrak{y}\, K[\underline{\underline{X}}] \subseteq \mathfrak{P}$, whence $\sqrt{\mathfrak{y}\, K[\underline{\underline{X}}]} \subseteq \sqrt{\mathfrak{P}} = \mathfrak{P}$.

<u>Example</u>: We give an example where $(\overline{\underline{\underline{x}}})^K = (\overline{\underline{\underline{x}}})^k$, but $\mathfrak{P} \neq \mathfrak{y}\, K[\underline{\underline{X}}]$. Thus (iii) does not imply (ii). Let k_0 be a field of characteristic p, and let $k = k_0(z)$, where z is transcendental over k_0. Put $\underline{\underline{x}} = (t, t\sqrt[p]{z})$, where t is transcendental over k. Then $\mathfrak{y} = \mathfrak{I}_k(\underline{\underline{x}}) = (zX_1^p - X_2^p)$, since $zX_1^p - X_2^p$ is an irreducible polynomial over k. Now take $K = k(\sqrt[p]{z})$. Then $\mathfrak{P} = \mathfrak{I}_K(\underline{\underline{x}}) = (\sqrt[p]{z}\, X_1 - X_2)$, and $\mathfrak{P} \neq \mathfrak{y}\, K[\underline{\underline{X}}]$. We have $(\overline{\underline{\underline{x}}})^k = A((zX_1^p - X_2^p))$ and $(\overline{\underline{\underline{x}}})^K = A((\sqrt[p]{z}\, X_1 - X_2))$. We observe that $(\overline{\underline{\underline{x}}})^k = (\overline{\underline{\underline{x}}})^K$, since if $(u,v) \in A((zX_1^p - X_2^p))$, then $zu^p - v^p = (\sqrt[p]{z}\, u - v)^p = 0$, so that $(u,v) \in A((\sqrt[p]{z}\, X_1 - X_2))$.

<u>THEOREM 6B</u>. <u>Let</u> k, K, $\underline{\underline{x}}$, $\mathfrak{y}$, $\mathfrak{P}$ <u>be as in Theorem 6A</u>. <u>Suppose, moreover, that</u> K <u>is a separable algebraic extension of</u> k. <u>Then</u> $\sqrt{\mathfrak{y}\, K[\underline{\underline{X}}]} = \mathfrak{y}\, K[\underline{\underline{X}}]$.

---

† Linearly independent vectors in a vector space k^t over k remain linearly independent in the vector space K^t, where K is an over-field of k.

<u>Proof</u>: Let $f \in \sqrt{\mathcal{Y} K[\underline{X}]}$. There is a field K_0 with $k \subseteq K_0 \subseteq K$ which is finitely generated over k , such that $f \in K_0[\underline{X}]$ and $f \in \sqrt{\mathcal{Y} K_0[\underline{X}]}$. Let $f = \sum_{i=1}^{n} c_i f_i$, where $f_i(\underline{X}) \in k[\underline{X}]$, $c_i \in K_0$, and $c_1,\ldots,c_n$ are linearly independent over k . In fact, by allowing some f_i to be zero, we may suppose that $c_1,\ldots,c_n$ are a basis for K_0 over k , where $n = [K_0 : k]$. There are n distinct k-isomorphisms σ of K_0 into Ω ; write c^{σ} for the image of c under σ . We put

$$f^{\sigma}(\underline{X}) = \sum_{i=1}^{n} c_i^{\sigma} f_i(\underline{X}) .$$

Here the $(n \times n)$-determinant $|c_i^{\sigma}|$ is not zero, and hence there are $d_i^{(\sigma)}$ such that

$$f_i(\underline{X}) = \sum_{\sigma} d_i^{(\sigma)} f^{\sigma}(\underline{X}) \qquad (i = 1,\ldots,n).$$

Now for some m , $f^m \in \mathcal{Y} K_0[\underline{X}]$, whence $(f^{\sigma})^m \in \mathcal{Y} K_0^{\sigma}[\underline{X}]$, whence $(f^{\sigma})^m(\underline{x}) = 0$, and therefore $f^{\sigma}(\underline{x}) = 0$ for each σ . Thus each $f_i(\underline{x}) = 0$, and $f_i \in \mathcal{Y}$. We have shown that $f \in \mathcal{Y} K_0[\underline{X}] \subseteq \mathcal{Y} K[\underline{X}]$.

It follows from Theorems 6A, 6B, that the four properties listed in Theorem 6A are equivalent if K is a separable algebraic extension of k . Now if k is perfect, then every algebraic extension K of k is separable. Thus we obtain

<u>COROLLARY 6C</u>. *If k is perfect and if V is a variety over k with generic point $\underline{x}$, then V is an absolute variety if and only*

if $k(\underline{x})$ *and* k *are linearly disjoint over* $\underline{k}$. *This is the case if and only if* k *is algebraically closed in* $k(\underline{x})$.[*)]

THEOREM 6D. Let k *be a perfect ground field.*

(i) *If* $f(\underline{X}) \in k[\underline{X}]$ *is not constant and is absolutely irreducible, then the set of zeros of* f *is an absolute hypersurface.*

(ii) *If* S *is an absolute hypersurface, then* $\mathfrak{J}_k(S) = (f)_k$[†], *where* f *is absolutely irreducible and nonconstant.*

Proof: (i) This follows directly from Theorem 2C, and the fact that f is absolutely irreducible.

(ii) From Theorem 2C it follows that $\mathfrak{J}_k(S) = (f)_k$, where f is nonconstant and irreducible over k. Let K be an algebraic extension of k. Then $\mathfrak{J}_K(S) = \mathfrak{P} = \mathfrak{p}K[\underline{X}] = (f)_k K[\underline{X}] = (f)_K$. Thus the principal ideal generated by f in $K[\underline{X}]$ is a prime ideal, and f is irreducible over K.

REMARKS (1). Let k be perfect and let V be a variety over k. In Theorem 4D we constructed a hypersurface S which was birationally equivalent to V. In fact, the construction was such that $k(\underline{x}) = k(\underline{y})$, where $\underline{x}$, $\underline{y}$ were certain generic points of V, S, respectively. Now if V is an absolute variety, then k is algebraically

---

[†] We write $(f)_k$ resp. $(f)_K$ for the principal ideal generated by f in $k[\underline{X}]$ and in $K[\underline{X}]$.

*) Compare with Theorem 3A of Ch. V.

closed in $k(\underline{\underline{x}}) = k(\underline{\underline{y}})$, and S is _also an absolute variety_.

(2) Another approach to Corollary 6C is this: It may be shown directly that if two k-varieties are k-birationally equivalent, and if one is absolute, then so is the other. Thus the proof may be reduced to the case of a hypersurface. But this case is essentially Theorem 3A of Ch. V.

7. Counting Points in Varieties Over Finite Fields

The goal of this section is a proof of

THEOREM 7A. _Let_ V _be an absolute variety of dimension_ d _defined over_ $k = F_q$. _Let_ $N_\nu = N_\nu(V)$ _be the number of points_ $\underline{\underline{y}} = (y_1,\ldots,y_n)$ _in_ V _with each coordinate in_ F_{q^ν}. _Then as_ $\nu \to \infty$,

$$N_\nu = q^{\nu d} + O\left(q^{\nu(d - 1/2)}\right) . \tag{7.1}$$

The proof will depend on a result we derived in Chapter V. Namely, if $f(X_1,\ldots,X_n) \in F_q[X_1,\ldots,X_n]$ is nonconstant and absolutely irreducible and if N is the number of zeros of f in F_q, then

$$|N - q^{n-1}| \leq cq^{n - 3/2} , \tag{7.2}$$

where c is a constant which depends on n and the total degree of f. For $n = 2$, this result is Theorem 1A of Chapter III, and for general n it is Theorem 5A of Chapter V. Only the case $n = 2$ is needed if V is a curve.

LEMMA 7B: Theorem 7A is true for hypersurfaces.

Proof: Let S be an absolute hypersurface of dimension d . By Theorem 6D, S is given by $f(\underline{\underline{x}}) = 0$, where $f(\underline{\underline{X}})$ is not constant and is absolutely irreducible. Thus by (7.2),

$$|N - q^d| = |N - q^{n-1}| \leq cq^{n-(3/2)} = cq^{d-1/2} .$$

Now applying this result to F_{q^ν} instead of F_q , we see that $|N_\nu - q^{\nu d}| \leq cq^{\nu(d-1/2)}$.

Theorem 7A for the general variety is done by induction on d . If $d = 0$ and $V = (\overline{\underline{\underline{x}}})$, then every $z \in F_q(\underline{\underline{x}})$ is algebraic over F_q , and so satisfies an equation $1 \cdot z - \alpha \cdot 1 = 0$ where $\alpha \in \overline{F}_q$. Thus z , 1 are linearly dependent over $\overline{F}_q$. Since $F_q(\underline{\underline{x}})$ and $\overline{F}_q$ are linearly disjoint over F_q , it follows that z , 1 are linearly dependent over F_q . So $z \in F_q$, and $F_q(\underline{\underline{x}}) = F_q$. Thus $\underline{\underline{x}}$ has coordinates in F_q , and $V = (\overline{\underline{\underline{x}}}) = \underline{\underline{x}}$. It follows that $N_\nu = 1$ for every ν .

In order to do the induction step from $d - 1$ to d , we shall need

LEMMA 7C. Suppose Theorem 7A is true for absolute varieties of dimension $< d$. Let W be a variety of dimension $< d$, not necessarily an absolute variety. Then as $\nu \to \infty$,

$$N_\nu(W) = O\left(q^{\nu(d-1)}\right) .$$

Proof: It is clear that W is still an algebraic set over $K = \overline{F}_q$, but not necessarily a K-variety. So W is a finite union $W = W_1 \cup \ldots \cup W_t$, where the W_i are K-varieties. Each W_i is defined by finitely many equations. The coefficients of all these equations for $W_1, \dots, W_t$ generate a finite extension F_{q^μ} of F_q. So each W_i is a F_{q^μ}-variety and is as such an absolute variety, and $d_i = \dim W_i \le d-1$. Let $N_{\lambda\mu}(W_i)$ be the number of points in W_i with coordinates in $F_{q^{\lambda\mu}}$. By our induction hypothesis, applied to F_{q^μ} instead of F_q, we see that as the integer λ tends to ∞, we have

$$\begin{aligned} N_{\lambda\mu}(W_i) &= q^{\lambda\mu(d_i-1)} + O\left(q^{\lambda\mu(d_i - 3/2)}\right) \\ &= O\left(q^{\lambda\mu(d-1)}\right) . \end{aligned}$$

Thus $N_{\lambda\mu}(W) = O\left(q^{\lambda\mu(d-1)}\right)$ as $\lambda \to \infty$. Given ν, pick an integer λ with $(\lambda-1)\mu < \nu \le \lambda\mu$. Then as $\nu \to \infty$,

$$\begin{aligned} N_\nu(W) \le N_{\lambda\mu}(W) &= O\left(q^{\lambda\mu(d-1)}\right) \\ &= O\left(q^{\nu(d-1) + \mu(d-1)}\right) \\ &= O\left(q^{\nu(d-1)}\right) . \end{aligned}$$

The proof of Theorem 7A is now completed as follows. According to Theorem 4D, the variety V is birationally equivalent to a hypersurface S, and this hypersurface is an absolute variety by the remark at the end of §6. By Theorem 4A, there exist proper algebraic subsets $L \subseteq V$, $M \subseteq S$, such that the birational correspondence $\underline{\underline{\omega}}$ between V and S becomes a $1-1$ correspondence between points of $V \sim L$ and of $S \sim M$. Now $\underline{\underline{\omega}}$ as well as its inverse is defined over $k = F_q$, i.e. is defined in terms of rational functions with coefficients in F_q. Thus in this correspondence, points with components in F_q correspond to points with components in F_q.

More generally, points with components in F_{q^ν} correspond to points with components in F_{q^ν} . Hence

(7.3) $$|N_\nu(V) - N_\nu(S)| \le N_\nu(L) + N_\nu(M) .$$

However, L and M are composed of varieties of dimension $< d$. So by Lemma 7C, $N_\nu(L) + N_\nu(M) = O\left(q^{\nu(d-1)}\right)$. On the other hand, by Lemma 7B, $N_\nu(S) = q^{\nu d} + O\left(q^{\nu(d - 1/2)}\right)$. These relations in conjunction with (7.3) yield (7.1).

REMARKS. (i) Theorem 7A together with Theorem 2D shows that the number N_ν of solutions $(x, y_1,\dots,y_t) \in F_{q^\nu}^{t+1}$ of certain systems of equations

$$y_1^{d_1} = g_1(x) , \quad y_2^{d_2} = g_2(x,y_1) ,\dots, \quad y_t^{d_t} = g_t(x, y_1,\dots, y_t)$$

satisfies $N_\nu = q^\nu + O(q^{\nu/2})$ as $\nu \to \infty$. In particular this holds for certain systems of equations

$$y_1^{d_1} = g_1(x),\dots, y_t^{d_t} = g_t(x) .$$

But a better result for such systems was already derived in Theorem 5A of Chapter II. Under suitable conditions on $g_1(X),\dots,g_t(X)$ it was shown that $|N_\nu - q^\nu| \leqq cq^{\nu/2}$, where c was a constant explicitly determined in terms of t and the degrees of the polynomials $g_1,\dots,g_t$.

(ii) More generally, if V is an absolute variety defined over F_q determined by equations $f_1(\underline{x}) = \dots = f_\ell(\underline{x}) = 0$, then our Theorem 7A could be strengthened to

$$|N_\nu - q^{\nu d}| \leqq cq^{\nu(d - 1/2)} ,$$

where c is a constant depending only on the number n of variables, on ℓ, and on the total degrees of the polynomials $f_1,\dots,f_t$.

(iii) Corollary 2B of Chapter V can be generalized as follows. Suppose V is an absolute variety of dimension d over $\mathbb{Q}$ defined by equations $f_1(\underline{x}) = \dots = f_\ell(\underline{x}) = 0$, where $f_1(\underline{X}),\dots,f_\ell(\underline{X})$ have rational integer coefficients. Let $\bar{f}_i(\underline{X})$ be obtained from $f_i(\underline{X})$ by reduction modulo p and let V_p be the algebraic set defined over F_p by $\bar{f}_1(\underline{x}) = \dots = \bar{f}_\ell(\underline{x}) = 0$. Then if $p > p_0$, the set V_p is an absolute variety of dimension d. Here p_0 depends only on n, ℓ and the degrees of the polynomials $f_1,\dots,f_\ell$. Hence if $p > p_0$, then the number $N(p)$ of solutions of the system of congruences

$$f_1(\underline{x}) \equiv \dots \equiv f_\ell(\underline{x}) \equiv 0 \pmod{p}$$

satisfies $|N(p)-p^d| \leqq cp^{d-1/2}$.

(iv) The Weil (1949) conjectures (see also Ch. IV, §6) imply much better estimates than Theorem 7A if V is a "non-singular" variety of dimension $d > 1$. These conjectures were recently proved by Deligne

---

+) But see the remark in the Preface.

BIBLIOGRAPHY

E. Artin (1924). Quadratische Körper im Gebiet der höheren Kongruenzen I, II. Math. Z. 19, 153-246.

(1955). Elements of Algebraic Geometry. Lecture Notes, New York Univ., Inst. of Math. Sciences.

J. Ax (1964). Zeros of polynomials over finite fields. Amer. J. Math. 86, 255-261.

E. Bertini (1882). Rendiconti R. Ist. Lombardo 15, 24-28.

A. S. Besicovitch (1940). On the linear independence of fractional powers of integers. J. London Math. Soc. 15, 3-6.

E. Bombieri (1966). On Exponential Sums in Finite Fields. Am. J. Math. 88, 71-105.

(1973). Counting points on curves over finite fields (d'après S. A. Stepanov). Seminaire Bourbaki, 25e année 1972/73, No. 430, Juin 1973.

Z. I. Borevich and I. R. Shafarevich (1966). Number theory. Academic Press. (Translated from the 1964 Russian Ed.)

L. Carlitz (1969). Kloosterman sums and finite field extensions. Acta Arith. 16, 179-193.

L. Carlitz and S. Uchiyama (1957). Bounds for exponential sums. Duke Math. J. 24, 37-41.

J. H. H. Chalk and R. A. Smith (1971). On Bombieri's estimate for exponential sums. Acta Arith. 18, 191-212.

C. Chevalley (1935). Démonstration d'une hypothèse de M. Artin. Abh. Math. Sem. Hamburg 11, 73-75.

H. Davenport and H. Hasse (1935). Nullstellen der Kongruenzzetafunktionen in gewissen zyklischen Fällen. J. Reine Ang. Math. 172, 151-182.

P. Deligne (1973). La conjecture de Weil. I.Inst. des Hautes Etudes Scientifiques Pub. Math. No. 48, 273-308.

M. Deuring (1958). Lectures on the theory of algebraic functions of one variable. (Tata Institute) (1973 Springer lecture notes No. 314).

B. Dwork (1960). On the rationality of the zeta function of an algebraic variety. Am. J. Math. 82, 631-648.

M. Eichler (1963). Einführung in die Theorie der algebraischen Zahlen und Funktionen. Birkhäuser Verlag.

C. F. Gauss (1801). Disquisitiones Arithmeticae. Fischer Verlag, Leipzig.

H. Hasse (1936a). Theorie der höheren Differentiale in einem algebraischen Funktionenkörper mit vollkommenem Konstantenkörper bei beliebiger Charakteristik. J. Reine Ang. Math. 175, 50-54.

(1936b). Zur Theorie der abstrakten elliptischen Funktionenkörper -- II. J. Reine Ang. Math. 175, 69-88.

(1936c). Zur Theorie der abstrakten elliptischen Funktionenkörper -- III. Ibid. 193-208.

D. Hilbert (1892). Über die Irreduzibilität ganzer rationaler Funktionen mit ganzzahligen Koeffizienten. J. Reine Ang. Math. 110, 104-129.

J. R. Joly (1973). Equations et variétés algébriques sur un corps fini. L'Enseignement Math. (2) 19, 1-113.

S. Lang (1958). Introduction to algebraic geometry. Interscience, New York - London.

(1961). Diophantine Geometry. Interscience, New York - London.

S. Lang and A. Weil (1954). The number of points of varieties in finite fields. Am. J. Math. 76, 819-827.

D. A. Mitkin (1972). On the estimation of a rational trigonometric sum with a prime denominator (In Russian). Vestnik Moscow Univ. (Mat., Mech.) 5, 50-58.

D. Mumford ( ). Lecture Notes on Algebraic Geometry. Harvard University.

L. B. Nisnevich (1954). On the number of points of an algebraic variety in a finite prime field. (In Russian). Dokl. Akad. Nauk SSSR 99, 17-20.

E. Noether (1922). Ein algebraisches Kriterium für absolute Irreduzibilität. Math. Ann. 85, 26-33.

A. Ostrowski (1919). Zur arithmetischen Theorie der algebraischen Größen. Göttinger Nachr. 279-298.

G. I. Perelmuter (1962). On certain character sums. (In Russian). Dokl. Akad. Nauk SSSR 144, 58-61.

G. I. Perelmuter and A. G. Postnikov (1972). On the number of solutions of monic equations. (In Russian). Acta Arith. 21, 103-110.

A. G. Postnikov (1967). Ergodic Aspects of the theory of congruences and of the theory of Diophantine Approximation. Transl. from the 1966 Russian ed. (Steklov Inst. of Math. No. 82). Am. Math. Soc., Providence, R. I.

P. Roquette (1953). Arithmetischer Beweis der Riemannschen Vermutung in Kongruenzzetafunktionenkörpern beliebigen Geschlechts. J. Reine Ang. Math. 191, 199-252.

W. M. Schmidt (1973). Zur Methode von Stepanov. Acta Arith. 24, 347-367.

(1974). A Lower Bound for the Number of Solutions of Equations over Finite Fields. J. Number Theory 6, 448-480.

I. R. Shafarevich (1974). Basic Algebraic Geometry (Transl. from the 1972 Russian Ed.) Springer Grundlehren, 213.

H. M. Stark (1973). On the Riemann Hypothesis in hyperelliptic function fields. AMS, Proc. of Symposia in Pure Math. 24, 285-302.

S. A. Stepanov (1969). The number of points of a hyperelliptic curve over a prime field (In Russian). Izv. Akad. Nauk SSSR Ser. Mat. 33, 1171-1181.

(1970). Elementary method in the theory of congruences for a prime modulus. Acta Arith. 17, 231-247.

(1971). Estimates of rational trigonometric sums with prime denominators (In Russian). Trudy Akad. Nauk 62, 346-371.

(1972a). An elementary proof of the Hasse-Weil Theorem for hyperelliptic curves. J. Number Theory 4, 118-143.

(1972b). Congruences in two variables. (In Russian). Izv. Akad. Nauk SSSR, Ser. Mat. 36, 683-711.

(1974). Rational points on algebraic curves over finite fields (In Russian). Report of a 1972 conference on analytic number theory in Minsk, USSR, 223-243.

O. Teichmüller (1936). Differentialrechnung bei Charakteristik p. J. Reine Ang. Math. 175, 89-99.

A. Thue (1909). Über Annäherungswerte algebraischer Zahlen. J. Reine Ang. Math. 135, 284-305.

B. L. Van der Waerden (1955). Algebra I, II (3rd ed.). Springer-Verlag. Berlin-Göttingen-Heidelberg.

E. Warning (1935). Bemerkung zur vorstehenden Arbeit von Herrn Chevalley. Abh. Math. Sem. Hamburg 11, 76-83.

A. Weil (1940). Sur les fonctions algébriques à corps de constantes fini. C. R. Acad. Sci. Paris 210, 592-594.

(1948a). Sur les courbes algébriques et les variétés qui s'en déduisent. Actualités sci. et ind. No. 1041.

(1948b). On some exponential sums. Proc. Nat. Acad. Sci. USA 34, 204-207.

(1949). Solutions of equations in finite fields. Bull. Am. Math. Soc. 55, 497-508.

O. Zariski (1941). Pencils on an algebraic variety and a new proof of a theorem of Bertini. Trans. A.M.S. 50, 48-70.

O. Zariski and P. Samuel (1958). Commutative Algebra I, II. D. Van Nostrand Company. Princeton - New York - London - Toronto.

Vol. 399: Functional Analysis and its Applications. Proceedings 1973. Edited by H. G. Garnir, K. R. Unni and J. H. Williamson. II, 584 pages. 1974.

Vol. 400: A Crash Course on Kleinian Groups. Proceedings 1974. Edited by L. Bers and I. Kra. VII, 130 pages. 1974.

Vol. 401: M. F. Atiyah, Elliptic Operators and Compact Groups. V, 93 pages. 1974.

Vol. 402: M. Waldschmidt, Nombres Transcendants. VIII, 277 pages. 1974.

Vol. 403: Combinatorial Mathematics. Proceedings 1972. Edited by D. A. Holton. VIII, 148 pages. 1974.

Vol. 404: Théorie du Potentiel et Analyse Harmonique. Edité par J. Faraut. V, 245 pages. 1974.

Vol. 405: K. J. Devlin and H. Johnsbråten, The Souslin Problem. VIII, 132 pages. 1974.

Vol. 406: Graphs and Combinatorics. Proceedings 1973. Edited by R. A. Bari and F. Harary. VIII, 355 pages. 1974.

Vol. 407: P. Berthelot, Cohomologie Cristalline des Schémas de Caracteristique p > o. II, 604 pages. 1974.

Vol. 408: J. Wermer, Potential Theory. VIII, 146 pages. 1974.

Vol. 409: Fonctions de Plusieurs Variables Complexes, Séminaire François Norguet 1970–1973. XIII, 612 pages. 1974.

Vol. 410: Séminaire Pierre Lelong (Analyse) Année 1972–1973. VI, 181 pages. 1974.

Vol. 411: Hypergraph Seminar. Ohio State University, 1972. Edited by C. Berge and D. Ray-Chaudhuri. IX, 287 pages. 1974.

Vol. 412: Classification of Algebraic Varieties and Compact Complex Manifolds. Proceedings 1974. Edited by H. Popp. V, 333 pages. 1974.

Vol. 413: M. Bruneau, Variation Totale d'une Fonction. XIV, 332 pages. 1974.

Vol. 414: T. Kambayashi, M. Miyanishi and M. Takeuchi, Unipotent Algebraic Groups. VI, 165 pages. 1974.

Vol. 415: Ordinary and Partial Differential Equations. Proceedings 1974. XVII, 447 pages. 1974.

Vol. 416: M. E. Taylor, Pseudo Differential Operators. IV, 155 pages. 1974.

Vol. 417: H. H. Keller, Differential Calculus in Locally Convex Spaces. XVI, 131 pages. 1974.

Vol. 418: Localization in Group Theory and Homotopy Theory and Related Topics. Battelle Seattle 1974 Seminar. Edited by P. J. Hilton. VI, 172 pages 1974.

Vol. 419: Topics in Analysis. Proceedings 1970. Edited by O. E. Lehto, I. S. Louhivaara, and R. H. Nevanlinna. XIII, 392 pages. 1974.

Vol. 420: Category Seminar. Proceedings 1972/73. Edited by G. M. Kelly. VI, 375 pages. 1974.

Vol. 421: V. Poénaru, Groupes Discrets. VI, 216 pages. 1974.

Vol. 422: J.-M. Lemaire, Algèbres Connexes et Homologie des Espaces de Lacets. XIV, 133 pages. 1974.

Vol. 423: S. S. Abhyankar and A. M. Sathaye, Geometric Theory of Algebraic Space Curves. XIV, 302 pages. 1974.

Vol. 424: L. Weiss and J. Wolfowitz, Maximum Probability Estimators and Related Topics. V, 106 pages. 1974.

Vol. 425: P. R. Chernoff and J. E. Marsden, Properties of Infinite Dimensional Hamiltonian Systems. IV, 160 pages. 1974.

Vol. 426: M. L. Silverstein, Symmetric Markov Processes. X, 287 pages. 1974.

Vol. 427: H. Omori, Infinite Dimensional Lie Transformation Groups. XII, 149 pages. 1974.

Vol. 428: Algebraic and Geometrical Methods in Topology, Proceedings 1973. Edited by L. F. McAuley. XI, 280 pages. 1974.

Vol. 429: L. Cohn, Analytic Theory of the Harish-Chandra C-Function. III, 154 pages. 1974.

Vol. 430: Constructive and Computational Methods for Differential and Integral Equations. Proceedings 1974. Edited by D. L. Colton and R. P. Gilbert. VII, 476 pages. 1974.

Vol. 431: Séminaire Bourbaki – vol. 1973/74. Exposés 436–452. IV, 347 pages. 1975.

Vol. 432: R. P. Pflug, Holomorphiegebiete, pseudokonvexe Gebiete und das Levi-Problem. VI, 210 Seiten. 1975.

Vol. 433: W. G. Faris, Self-Adjoint Operators. VII, 115 pages. 1975.

Vol. 434: P. Brenner, V. Thomée, and L. B. Wahlbin, Besov Spaces and Applications to Difference Methods for Initial Value Problems. II, 154 pages. 1975.

Vol. 435: C. F. Dunkl and D. E. Ramirez, Representations of Commutative Semitopological Semigroups. VI, 181 pages. 1975.

Vol. 436: L. Auslander and R. Tolimieri, Abelian Harmonic Analysis, Theta Functions and Function Algebras on a Nilmanifold. V, 99 pages. 1975.

Vol. 437: D. W. Masser, Elliptic Functions and Transcendence. XIV, 143 pages. 1975.

Vol. 438: Geometric Topology. Proceedings 1974. Edited by L. C. Glaser and T. B. Rushing. X, 459 pages. 1975.

Vol. 439: K. Ueno, Classification Theory of Algebraic Varieties and Compact Complex Spaces. XIX, 278 pages. 1975

Vol. 440: R. K. Getoor, Markov Processes: Ray Processes and Right Processes. V, 118 pages. 1975.

Vol. 441: N. Jacobson, PI-Algebras. An Introduction. V, 115 pages. 1975.

Vol. 442: C. H. Wilcox, Scattering Theory for the d'Alembert Equation in Exterior Domains. III, 184 pages. 1975.

Vol. 443: M. Lazard, Commutative Formal Groups. II, 236 pages. 1975.

Vol. 444: F. van Oystaeyen, Prime Spectra in Non-Commutative Algebra. V, 128 pages. 1975.

Vol. 445: Model Theory and Topoi. Edited by F. W. Lawvere, C. Maurer, and G. C. Wraith. III, 354 pages. 1975.

Vol. 446: Partial Differential Equations and Related Topics. Proceedings 1974. Edited by J. A. Goldstein. IV, 389 pages. 1975.

Vol. 447: S. Toledo, Tableau Systems for First Order Number Theory and Certain Higher Order Theories. III, 339 pages. 1975.

Vol. 448: Spectral Theory and Differential Equations. Proceedings 1974. Edited by W. N. Everitt. XII, 321 pages. 1975.

Vol. 449: Hyperfunctions and Theoretical Physics. Proceedings 1973. Edited by F. Pham. IV, 218 pages. 1975.

Vol. 450: Algebra and Logic. Proceedings 1974. Edited by J. N. Crossley. VIII, 307 pages. 1975.

Vol. 451: Probabilistic Methods in Differential Equations. Proceedings 1974. Edited by M. A. Pinsky. VII, 162 pages. 1975.

Vol. 452: Combinatorial Mathematics III. Proceedings 1974. Edited by Anne Penfold Street and W. D. Wallis. IX, 233 pages. 1975.

Vol. 453: Logic Colloquium. Symposium on Logic Held at Boston, 1972–73. Edited by R. Parikh. IV, 251 pages. 1975.

Vol. 454: J. Hirschfeld and W. H. Wheeler, Forcing, Arithmetic, Division Rings. VII, 266 pages. 1975.

Vol. 455: H. Kraft, Kommutative algebraische Gruppen und Ringe. III, 163 Seiten. 1975.

Vol. 456: R. M. Fossum, P. A. Griffith, and I. Reiten, Trivial Extensions of Abelian Categories. Homological Algebra of Trivial Extensions of Abelian Categories with Applications to Ring Theory. XI, 122 pages. 1975.

Vol. 457: Fractional Calculus and Its Applications. Proceedings 1974. Edited by B. Ross. VI, 381 pages. 1975.

Vol. 458: P. Walters, Ergodic Theory – Introductory Lectures. VI, 198 pages. 1975.

Vol. 459: Fourier Integral Operators and Partial Differential Equations. Proceedings 1974. Edited by J. Chazarain. VI, 372 pages. 1975.

Vol. 460: O. Loos, Jordan Pairs. XVI, 218 pages. 1975.

Vol. 461: Computational Mechanics. Proceedings 1974. Edited by J. T. Oden. VII, 328 pages. 1975.

Vol. 462: P. Gérardin, Construction de Séries Discrètes p-adiques. »Sur les séries discrètes non ramifiées des groupes réductifs déployés p-adiques«. III, 180 pages. 1975.

Vol. 463: H.-H. Kuo, Gaussian Measures in Banach Spaces. VI, 224 pages. 1975.

Vol. 464: C. Rockland, Hypoellipticity and Eigenvalue Asymptotics. III, 171 pages. 1975.

Vol. 465: Séminaire de Probabilités IX. Proceedings 1973/74. Edité par P. A. Meyer. IV, 589 pages. 1975.

Vol. 466: Non-Commutative Harmonic Analysis. Proceedings 1974. Edited by J. Carmona, J. Dixmier and M. Vergne. VI, 231 pages. 1975.

Vol. 467: M. R. Essén, The Cos $\pi\lambda$ Theorem. With a paper by Christer Borell. VII, 112 pages. 1975.

Vol. 468: Dynamical Systems – Warwick 1974. Proceedings 1973/74. Edited by A. Manning. X, 405 pages. 1975.

Vol. 469: E. Binz, Continuous Convergence on C(X). IX, 140 pages. 1975.

Vol. 470: R. Bowen, Equilibrium States and the Ergodic Theory of Anosov Diffeomorphisms. III, 108 pages. 1975.

Vol. 471: R. S. Hamilton, Harmonic Maps of Manifolds with Boundary. III, 168 pages. 1975.

Vol. 472: Probability-Winter School. Proceedings 1975. Edited by Z. Ciesielski, K. Urbanik, and W. A. Woyczyński. VI, 283 pages. 1975.

Vol. 473: D. Burghelea, R. Lashof, and M. Rothenberg, Groups of Automorphisms of Manifolds. (with an appendix by E. Pedersen) VII, 156 pages. 1975.

Vol. 474: Séminaire Pierre Lelong (Analyse) Année 1973/74. Edité par P. Lelong. VI, 182 pages. 1975.

Vol. 475: Répartition Modulo 1. Actes du Colloque de Marseille-Luminy, 4 au 7 Juin 1974. Edité par G. Rauzy. V, 258 pages. 1975. 1975.

Vol. 476: Modular Functions of One Variable IV. Proceedings 1972. Edited by B. J. Birch and W. Kuyk. V, 151 pages. 1975.

Vol. 477: Optimization and Optimal Control. Proceedings 1974. Edited by R. Bulirsch, W. Oettli, and J. Stoer. VII, 294 pages. 1975.

Vol. 478: G. Schober, Univalent Functions – Selected Topics. V, 200 pages. 1975.

Vol. 479: S. D. Fisher and J. W. Jerome, Minimum Norm Extremals in Function Spaces. With Applications to Classical and Modern Analysis. VIII, 209 pages. 1975.

Vol. 480: X. M. Fernique, J. P. Conze et J. Gani, Ecole d'Eté de Probabilités de Saint-Flour IV–1974. Edité par P.-L. Hennequin. XI, 293 pages. 1975.

Vol. 481: M. de Guzmán, Differentiation of Integrals in R^n. XII, 226 pages. 1975.

Vol. 482: Fonctions de Plusieurs Variables Complexes II. Séminaire François Norguet 1974–1975. IX, 367 pages. 1975.

Vol. 483: R. D. M. Accola, Riemann Surfaces, Theta Functions, and Abelian Automorphisms Groups. III, 105 pages. 1975.

Vol. 484: Differential Topology and Geometry. Proceedings 1974. Edited by G. P. Joubert, R. P. Moussu, and R. H. Roussarie. IX, 287 pages. 1975.

Vol. 485: J. Diestel, Geometry of Banach Spaces – Selected Topics. XI, 282 pages. 1975.

Vol. 486: S. Stratila and D. Voiculescu, Representations of AF-Algebras and of the Group U (∞). IX, 169 pages. 1975.

Vol. 487: H. M. Reimann und T. Rychener, Funktionen beschränkter mittlerer Oszillation. VI, 141 Seiten. 1975.

Vol. 488: Representations of Algebras, Ottawa 1974. Proceedings 1974. Edited by V. Dlab and P. Gabriel. XII, 378 pages. 1975.

Vol. 489: J. Bair and R. Fourneau, Etude Géométrique des Espaces Vectoriels. Une Introduction. VII, 185 pages. 1975.

Vol. 490: The Geometry of Metric and Linear Spaces. Proceedings 1974. Edited by L. M. Kelly. X, 244 pages. 1975.

Vol. 491: K. A. Broughan, Invariants for Real-Generated Uniform Topological and Algebraic Categories. X, 197 pages. 1975.

Vol. 492: Infinitary Logic: In Memoriam Carol Karp. Edited by D. W. Kueker. VI, 206 pages. 1975.

Vol. 493: F. W. Kamber and P. Tondeur, Foliated Bundles and Characteristic Classes. XIII, 208 pages. 1975.

Vol. 494: A Cornea and G. Licea. Order and Potential Resolvent Families of Kernels. IV, 154 pages. 1975.

Vol. 495: A. Kerber, Representations of Permutation Groups II. V, 175 pages. 1975.

Vol. 496: L. H. Hodgkin and V. P. Snaith, Topics in K-Theory. Two Independent Contributions. III, 294 pages. 1975.

Vol. 497: Analyse Harmonique sur les Groupes de Lie. Proceedings 1973–75. Edité par P. Eymard et al. VI, 710 pages. 1975.

Vol. 498: Model Theory and Algebra. A Memorial Tribute to Abraham Robinson. Edited by D. H. Saracino and V. B. Weispfenning. X, 463 pages. 1975.

Vol. 499: Logic Conference, Kiel 1974. Proceedings. Edited by G. H. Müller, A. Oberschelp, and K. Potthoff. V, 651 pages 1975.

Vol. 500: Proof Theory Symposion, Kiel 1974. Proceedings. Edited by J. Diller and G. H. Müller. VIII, 383 pages. 1975.

Vol. 501: Spline Functions, Karlsruhe 1975. Proceedings. Edited by K. Böhmer, G. Meinardus, and W. Schempp. VI, 421 pages. 1976.

Vol. 502: János Galambos, Representations of Real Numbers by Infinite Series. VI, 146 pages. 1976.

Vol. 503: Applications of Methods of Functional Analysis to Problems in Mechanics. Proceedings 1975. Edited by P. Germain and B. Nayroles. XIX, 531 pages. 1976.

Vol. 504: S. Lang and H. F. Trotter, Frobenius Distributions in GL_2-Extensions. III, 274 pages. 1976.

Vol. 505: Advances in Complex Function Theory. Proceedings 1973/74. Edited by W. E. Kirwan and L. Zalcman. VIII, 203 pages. 1976.

Vol. 506: Numerical Analysis, Dundee 1975. Proceedings. Edited by G. A. Watson. X, 201 pages. 1976.

Vol. 507: M. C. Reed, Abstract Non-Linear Wave Equations. VI, 128 pages. 1976.

Vol. 508: E. Seneta, Regularly Varying Functions. V, 112 pages. 1976.

Vol. 509: D. E. Blair, Contact Manifolds in Riemannian Geometry. VI, 146 pages. 1976.

Vol. 510: V. Poènaru, Singularités C^∞ en Présence de Symétrie. V, 174 pages. 1976.

Vol. 511: Séminaire de Probabilités X. Proceedings 1974/75. Edité par P. A. Meyer. VI, 593 pages. 1976.

Vol. 512: Spaces of Analytic Functions, Kristiansand, Norway 1975. Proceedings. Edited by O. B. Bekken, B. K. Øksendal, and A. Stray. VIII, 204 pages. 1976.

Vol. 513: R. B. Warfield, Jr. Nilpotent Groups. VIII, 115 pages. 1976.

Vol. 514: Séminaire Bourbaki vol. 1974/75. Exposés 453 – 470. IV, 276 pages. 1976.

Vol. 515: Bäcklund Transformations. Nashville, Tennessee 1974. Proceedings. Edited by R. M. Miura. VIII, 295 pages. 1976.